TI-83 PLUS/TI-89 MANUAL

PATRICIA HUMPHREY

Georgia Southern University

STATS
MODELING THE WORLD

David E. Bock

Ithaca High School

Paul F. Velleman

Cornell University

Richard D. De Veaux

Williams College

Boston San Francisco New York
London Toronto Sydney Tokyo Singapore Madrid
Mexico City Munich Paris Cape Town Hong Kong Montreal

Reproduced by Pearson Addison-Wesley from electronic files supplied by the author.

Publishing as Pearson Addison-Wesley, 75 Arlington Street, Boston, MA 02116

ISBN 0-321-19406-3

1 2 3 4 5 6 VHG 06 05 04 03

Part 1 – Statistics with the TI-83

Part 2 – Statistics with the TI-89

Foreword

This calculator manual has been written to accompany *Stats: Modeling the World* by David E. Bock, Paul F. Velleman and Richard D. DeVeaux. Its chapters have been organized to closely follow the order of topic presentation in the text. The first half of the manual contains instructions and examples using the TI-83 calculator. The second half repeats the material using the TI-89.

If, as the authors of the test assert, and I concur with their statement, statistics is a process involving three main steps: Think, Show, and Tell; then this book is primarily about the Show step. Calculators (and computers) are in essence tools that help with the mechanics of a process. The end results they give are numbers or plots. They cannot decide what procedure is desirable or valid for any given set of data. It is up to the individual to decide if the assumptions and conditions for the desired analysis are fulfilled; if not, the analysis is invalid and any conclusions drawn are bogus. Similarly, the numbers given as results of a procedure are not the ending point. These must be interpreted in the context of the situation to draw appropriate conclusions.

In my examples, I have given full instructions in use of the calculators along with screen shots as appropriate to illustrate the inputs and outputs. I have also made appropriate conclusions as warranted. I have also included in each chapter, as appropriate, a What Can Go Wrong? Section which illustrates common error messages and errors and tells how to correct these. These generally occur where an error might first be encountered.

Chapter 1 - TI-83 Basics

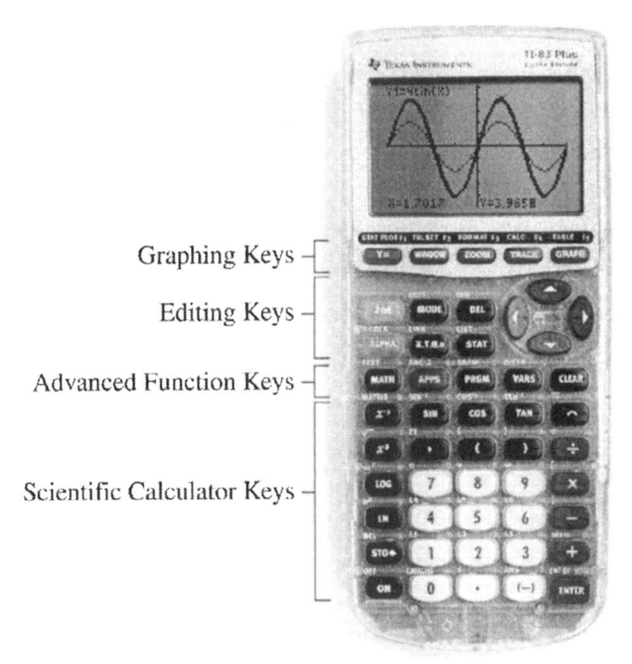

Figure 1.1 The TI-83 Plus Calculator

Both the TI-83 and the TI-83 Plus calculators include commonly used statistical functions, as well as most basic inferential methods. The major difference between the two is the addition of Flash memory in the 83 Plus and the availability of many downloadable applications.

Commonly Used Buttons and Commands

[ON]	At the lower left, this button turns the calculator on. When first turned on, the calculator displays the "home" screen.
[2nd]	The gold key on the upper left. This changes the cursor to ⬆ and is used to access items listed in gold above keys.
[2nd][ON]	Turns the calculator off. It will automatically turn itself off to save battery life after a few minutes of inactivity.
[ENTER]	The key on the lower right. This is used to execute commands.
[2nd] [ENTER]	Used to recall the last command entered. Used repeatedly, this cycles through the last several commands.
[CLEAR]	This key is on the upper right, just below the down arrow. It is used to clear commands on the home screen and lists in the Statistics Editor.
[2nd][MODE]	Executes the quit command, which leaves a menu or exits the Statistics and Matrix editors.

[STAT] Located just to the left of the left arrow key. It transfers to the initial Statistics menu.

[−] Located on the right side, just above the + key. This is the arithmetic "minus."

[(-)] Located next to [ENTER]. This is the "negative" key. Don't confuse it with the [−].

[ALPHA] This is the green key located at the upper left. This toggles the cursor between letters (indicated in green above keys) and standard key functions.

[STO▸] Located just above [ON], this allows storage of many results.

[APPS] On the TI-83 Plus, this allows access to Flash applications which include the Finance menu.

Cursor movement and Command Editing

[▲], [▼] Up and down arrow keys are used to move the cursor around menus.

[◄], [►] Left and right arrow keys will move the cursor one keystroke when editing commands. In graphics displays, with [TRACE]activated, they move the cursor around the plot.

[2nd][◄], [2nd][►] Moves the cursor to the beginning or end of a command.

[DEL] Located next to the left arrow key near the upper right, deletes the highlighted character in a command.

[2nd][DEL] This is the Insert command which allows characters to be inserted when editing.

Graphics keys are blue and are located on the very top row of the calculator.

[Y=] A blue key located on the upper left. It is used to enter functions.

[WINDOW] Sets graphing windows; this is usually only necessary for histogram plots.

[ZOOM] Accesses the basic windowing menu.

[TRACE] Allows moving the cursor around a graph, displaying coordinates.

[GRAPH] Displays a graph with the current window settings.

Adjusting the Display Contrast

[2nd][▲] Hold down the [▲] key to make the screen darker.

[2nd][▼] Hold down the [▼] key to make the screen lighter.

Menu Operation

Many keys on the calculator activate menus which also may have sub-menus. For example, pressing [STAT] displays this menu. There are three sub menus: EDIT, CALC, and TESTS. To transfer between sub-menus use the [▶] or [◀] keys.

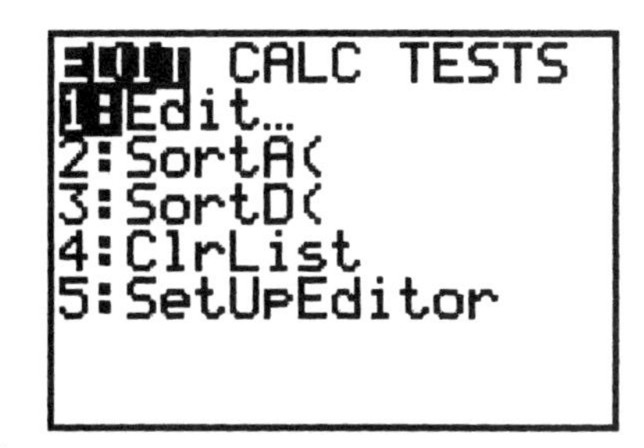

Selecting a menu option can be done either by pressing the [▼] or [▲] button until the desired choice is highlighted, then pressing [ENTER], or simply typing the number (or alpha character) of the menu option.

Sharing Data and Programs between Calculators

Data and programs may be shared between calculators using the communications cable which is supplied. The TI-83 and TI-83 Plus can share any TI-83 information and any TI-83 Plus information with the exception of applications and their associated variables.

Connect the supplied cable to the port at the base of each calculator. On both calculators press [2nd][X,T,Θ,n] to activate the LINK menu.

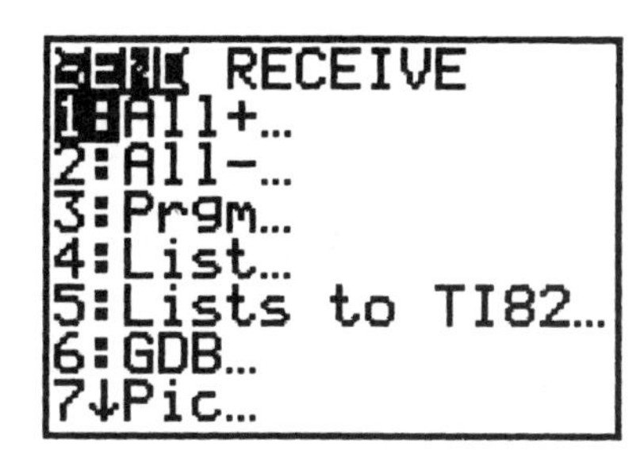

On the receiving calculator, press [▶] to highlight RECEIVE, then press [ENTER]. The calculator will display the message "Waiting..." The rolling cursor on the upper right indicates the calculator is working.

On the sending calculator, use the arrow keys to select the type of information to send. For example sending lists, either arrow to 4:List and press [ENTER] or press [4]. The screen at right will be shown.

```
SELECT TRANSMIT
▸ L1        LIST
  L2        LIST
  L3        LIST
  L4        LIST
  L5        LIST
  L6        LIST
  RESID     LIST
```

To select items to send, move the cursor to the item, press [ENTER] to select it. After selecting all items to send, press [▶] to highlight TRANSMIT, then press [ENTER].

Sorry, TI-83s cannot communicate with TI-89s.

Sharing Data between the Calculator and a Computer

Data lists, screen shots and programs may be shared between the calculator and either Microsoft Windows or Macintosh computers using a special cable and either TI-Connect or TI-Graph Link software. Cables are available for either serial or ISBN ports and can be found through many outlets such as OfficeMax, and Amazon.com. The software can be downloaded free through the Texas Instruments website at education.ti.com.

Working with Lists

The basic building blocks of any statistical analysis are lists of data. Before doing any statistics plot or analysis the data must be entered into the calculator. The calculator has six lists available in the statistics editor; these are L1 through L6. Accessing lists by name on the TI-83 is done by pressing [2nd] followed by the list number.

Accessing the Statistics Editor

On the TI-83, press [STAT], 1: Edit... will be highlighted. Press [ENTER] to select this function.

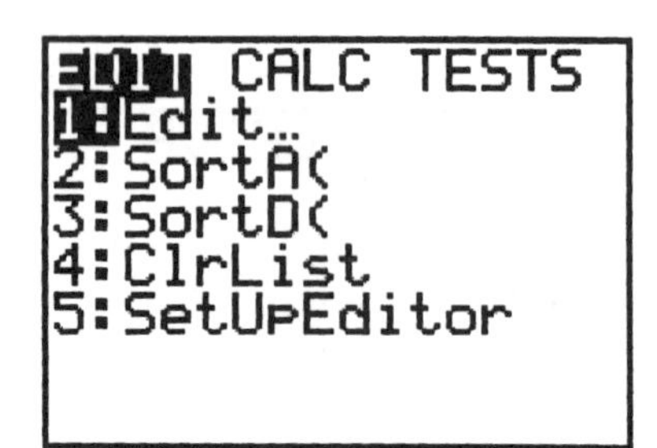

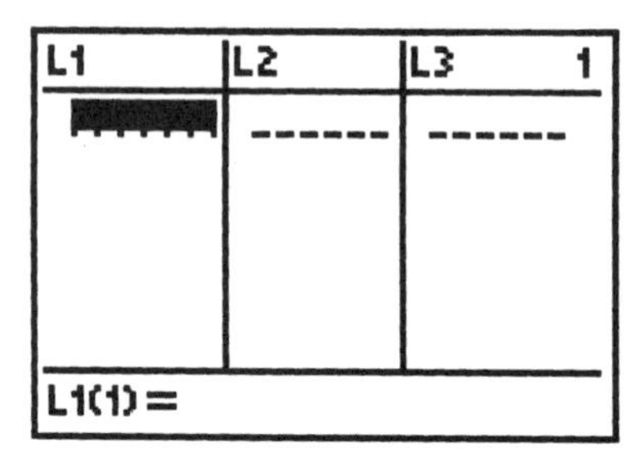

To enter data, simply use the right or left arrows to select a list, then type the entries in the list, following each value with [ENTER]. Note that it's not necessary to type any trailing zeros. They won't even be seen unless decimal places (found with the [MODE] key) have been set to some fixed number.

One word of advice: Most lists of data in texts are entered across the page in order to save space. Don't think that just because there are 4 (or more!) columns of data they belong in 4 (or more) lists. Data which belongs to a single variable always belongs in a single list.

Editing Data

It is always recommended that a list be double-checked for accuracy after entering it. Use the [▲] and [▼] keys to scroll through the list. If an entry is found to be in error, simply type the correction over the current value, then press [ENTER].

If a value needs to be inserted in the list, one can scroll to the bottom of the list and add it in the case of a single variable. However, if data are paired (such as in regression, frequency tables or paired tests) you will need to either delete the corresponding entry from the second list and place both at the bottom of each list or insert an item. Place the cursor on the value below where a new one is to be entered. Press [2nd][DEL] and a new line with a 0 value will be inserted. Move the cursor to the new value and type it in. In the example below, a new entry for the third element of the list is desired.

L1 L2 L3 1
1
12
15
9
16
19
L1(3)=17

L1 L2 L3 1
1
12
17
15
9
16
L1(3)=0

To delete an item, move the cursor to the desired item, then press [DEL]. In the example below, the fourth element in list L1 (17) is deleted.

L1 L2 L3 1
1
12
14
15
9
16
L1(4)=17

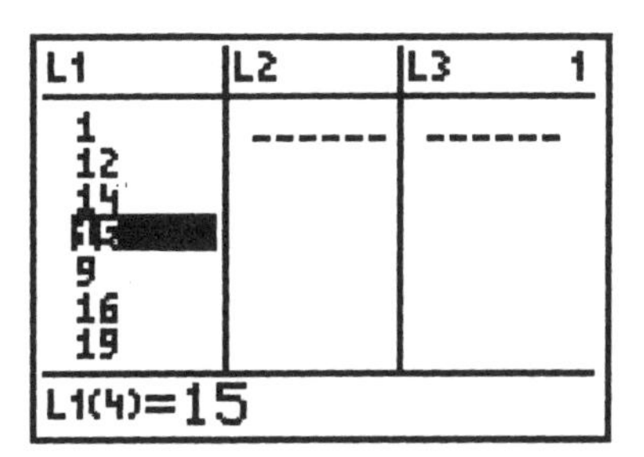

Exiting the Editor

To exit (leave) the statistics editor on the TI-83, press [2nd][MODE] (Quit). Note that it is not always necessary to leave the statistics editor before performing a new function. For example, if once a list is entered descriptive statistics are desired, press [STAT], arrow to CALC, press [ENTER] to select `1-Var Stats`, then input the list name ([2nd] followed by the number of the list) press [ENTER] to perform the calculation.)

Erasing Lists

There are several ways to erase (clear) all values in a list. Probably the easiest is to move the cursor so the name of the list is highlighted, then press [CLEAR][ENTER]. Never press [DEL] instead of [CLEAR].

Another way to clear lists either singly or several at once on the TI-83 is from the main STAT menu. This is choice `4:ClrList`. Press [STAT], either arrow to choice 4 and press [ENTER] or merely press [4]. The command is transferred to the home screen. Enter the list name(s) by pressing [2nd]n, where n is the number of the list. If more than one list is to be cleared at once, separate the list names with a comma. Press [ENTER] to execute the command. In the two examples below, the first clears list L1, the second lists L2 through L4.

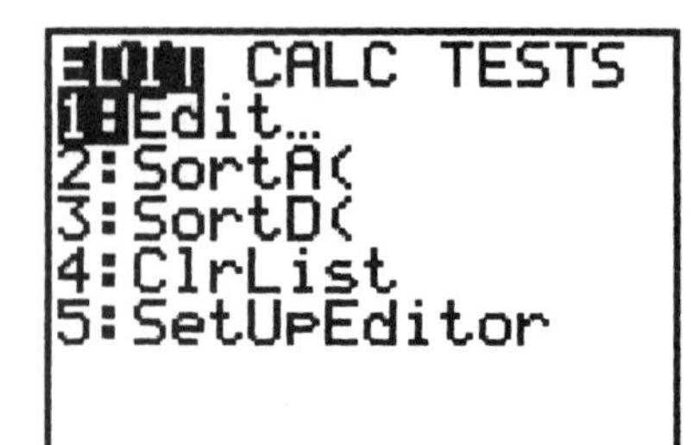

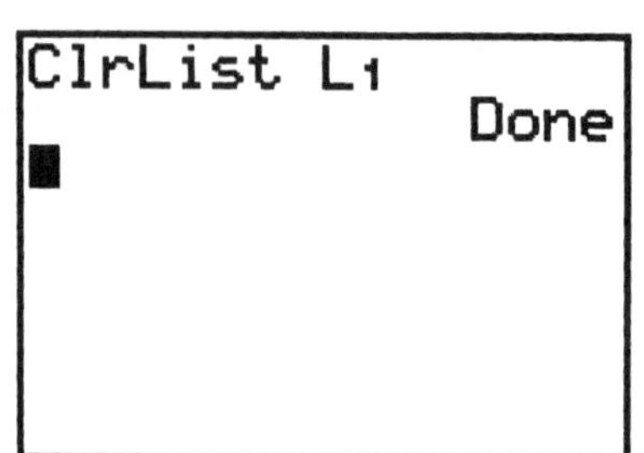

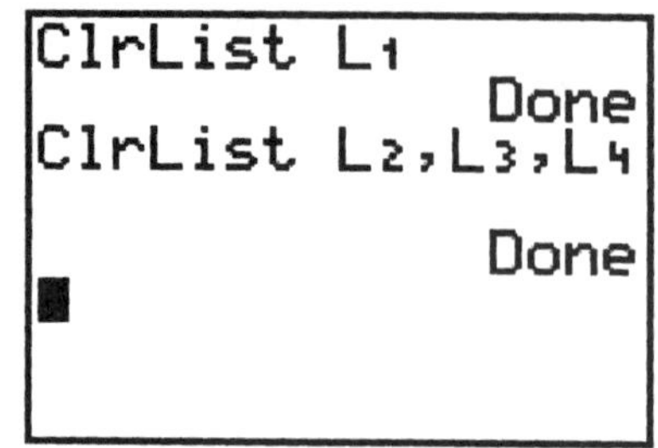

Sorting Lists

Lists may be sorted in either ascending (smallest to largest value) or descending order. The resulting list will replace the original list. For the TI-83, on the main [STAT] menu select either `2:SortA(` or `3:SortD(`. The command will be transferred to the home screen. Enter the name of the list to be sorted ([2nd], n where n is the number of the list). Execute the command by pressing [ENTER]. The example below sorts list L1.

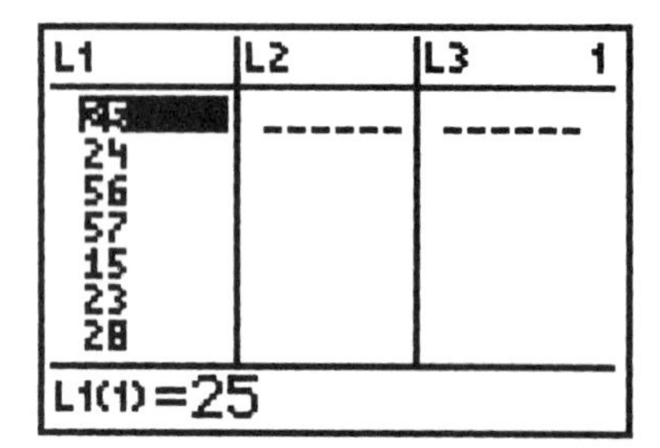

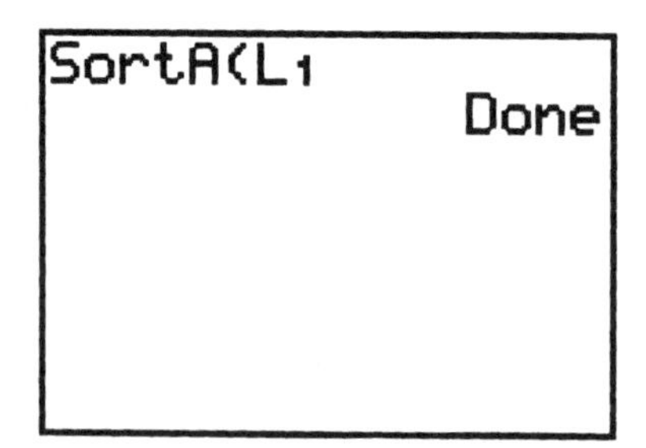

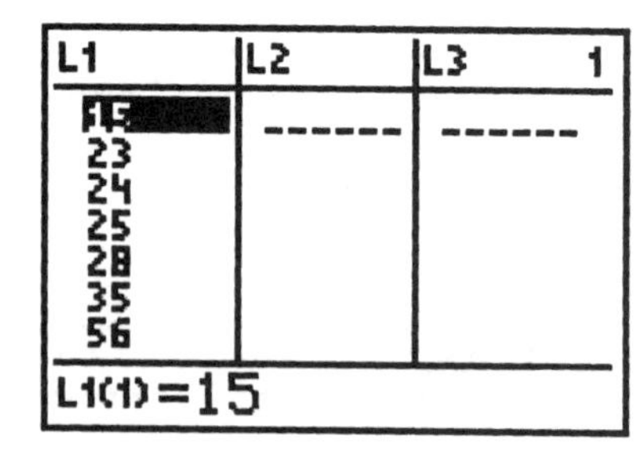

What Can Go Wrong?

Why is my list missing?

By far the most common error, aside from typographical errors is improper deletion of lists. When lists seem to be "missing" the user has pressed [DEL] rather than [CLEAR] in attempting to erase a list. Believe it or not, the data and the list are still in memory. To reclaim the missing list press [STAT] and select choice `5:SetUpEditor` followed by [ENTER] to execute the command. Upon return to the Editor, the missing list will be displayed.

Chapter 2 – Displaying Quantitative Data

The three primary rules of data analysis are:
1. Make a picture.
2. Make a picture.
3. Make a picture.

TI-calculators can help with this, although they cannot make bar graphs or pie charts for categorical data or dotplots and stem-and-leaf displays. All of these are fairly easily done by hand (at least for small data sets; for larger ones, use a full computer package).

This chapter will discuss histograms and time plots. Other plots of quantitative data will be discussed later. Histograms are examined for shape (skewed right/left or symmetric), center, and spread. They also tell us whether or not a distribution in unimodal (one-humped) or multi-modal (many humps). Time plots give indications of trends (systematic rising or falling patterns) and cycles (repeating patterns of rise and fall) in data which are observations on the same variable at different time points; they can also be used to examine the nature of volatility (how short or tall the peaks and valleys are).

In this chapter we will primarily work with the following data which represent monthly stock price changes in dollars for Enron stock for the period January 1997 to December 2001, just before the company collapsed. Looking at a table with lots of numbers is not a good way to understand what they show and to see patterns in the data.

	Jan	**Feb**	**Mar**	**Apr**	**May**	**Jun**	**Jul**	**Aug**	**Sep**	**Oct**	**Nov**	**Dec**
1997	-$1.44	-0.75	-0.69	-0.88	0.12	0.75	0.81	-1.75	0.69	-0.22	-0.16	0.34
1998	0.78	0.62	2.44	-0.28	2.22	-0.50	2.06	-0.88	-4.50	4.12	1.16	-0.50
1999	3.28	3.34	-1.22	0.47	5.62	-1.59	4.31	1.47	-0.72	-0.38	-3.25	0.03
2000	5.72	21.06	4.50	4.56	-1.25	-1.19	-3.12	8.00	9.31	1.12	-3.19	-17.75
2001	14.38	-1.08	-10.11	-12.11	5.84	-9.37	-4.74	-2.69	-10.61	-5.85	-17.16	-11.59

Histograms

Histograms are connected barcharts. Since the data are presumed to represent particular observations on some (underlying) continuous portion of the real number line and since order here matters, bars are always displayed connected to one another (unless there happens to be a gap in the values.) A good histogram has equal bar widths, high and low ends not too dramatically different from the maximum and minimum values of the data, and intervals which "make sense." As we will see, they are useful in featuring major features of the distribution of a single variable or for comparing two distributions (if done properly); they also have a capacity for "artiness" since their shape can change, depending on the choice of beginning values and bar widths.

L1	L2	L3 2
-1.44		------
-.75		
-.69		
-.88		
.12		
.75		
.81		

L2(1)=

The first step in making a histogram is to enter the data. The Enron data have been entered into list L1; the first few values are seen in the accompanying figure. Notice that it's not necessary to type all the leading 0's when doing the data entry. Also, be sure to lead the negatives with the [(-)] key not the [−] (minus) key.

TI-83 Steps to Create a Histogram

The next step is to define the plot. This is done by pressing [2nd][Y=] (Stat Plot). You will see the screen at right. Notice that there are three plots which can be displayed at any one time. For most purposes, there should be only one turned "on" at once. Notice here Plot1 is On and Plots 2 and 3 show Off. Scrolling down the menu are options 4 and 5 that turn all plots off or on with a single command. Selecting these will transfer the command to the home screen. Executing it requires pressing [ENTER].

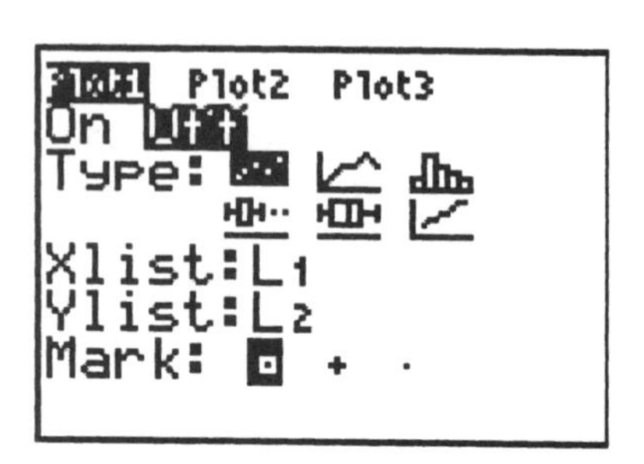

Press [ENTER] to select Plot1. The cursor should be blinking over the word On. If On is not already highlighted, press [ENTER] to move the highlight and to select displaying the plot. Notice there are six graphics types. Histograms are the third choice. Pressing [▼] will move the cursor to the first plot type, then use [▶] to move the cursor to the histogram figure. Press [ENTER] to move the highlight.

At this point, your screen probably looks like the one at right. We're ready to display the graph, since our data was in list L1 and each data value had frequency 1 (represented one occurrence of the value.) If you want to graph data in other lists, move the cursor to Xlist: and enter the list name ([2nd] n, where n is the number of the list). We'll talk more about frequencies later. Notice if you move the cursor to Freq: it will flash as 🄰. If you need to change this back from something else to a 1 you will need to press [ALPHA] before typing the 1.

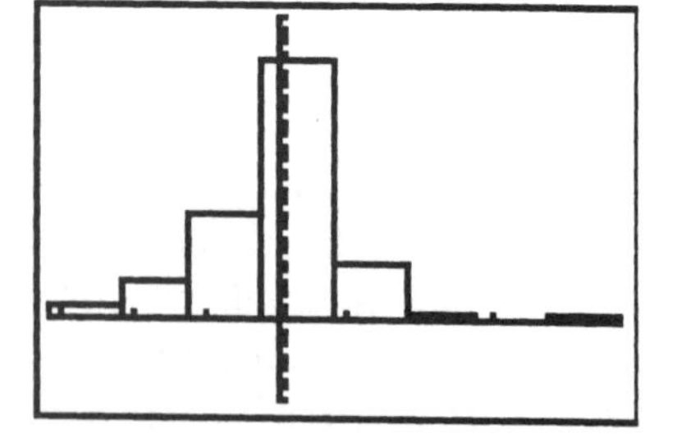

The easiest way to display a histogram (or any statistics plot) is to press [ZOOM][9] (Zoom Stat). The resulting graph is seen at right. Notice the *Y*-axis penetrating one of the bars. The *X*-axis "floats" a little way up from the bottom of the screen. This is so that values as seen in the next picture do not interfere with the plot.

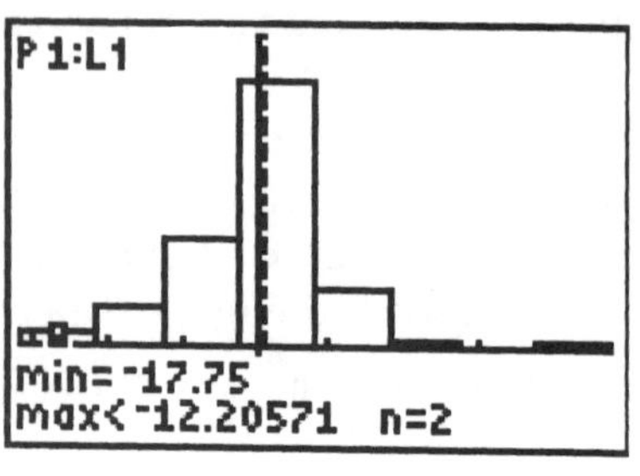

You will want to see just what the graph shows. To do this, press [TRACE]. A blinking cursor will show in the first bar at the left of the graph. At the bottom of the screen the minimum value included in the bar, maximum value for the bar, and number of observations in the bar will be displayed. This bar goes from -$17.75 to -$12.20571. There are two observations in this interval. Pressing the right arrow key ([▶]) will allow you to continue through the graph seeing the interval ranges and numbers of observations.

At this point, we can see the distribution of Enron stock prices appears to be unimodal and relatively symmetric (bars fall roughly equally from the center peak.) We see the center is around a price change of $0 with 31 observations in that bar; the worst change was a loss of $17.75 and the biggest gain was less than about $26.61. There may be an outlier at the high end – there is a gap in the histogram on the right hand side.

There is a downfall to using simply [ZOOM][9] for histograms. Look at the interval. It doesn't really make sense in a natural way. The bar width represents a difference between the low and high ends of each bar of $5.54428… which is unnatural. We'd like to fix this.

Manipulating Windows

To force particular minimums, maximums and scaling we will press [WINDOW]. This displays the screen at right. Notice the Xmin was the smallest value shown on the plot and Xscl was the bar width. These are quantities we'd like to change. You generally won't have to change any of the Y variables here (unless a scaling change loses the top of a bar – then increase Ymax). Another reason to possibly change Y variables is to increase resolution.

```
WINDOW
 Xmin=-17.75
 Xmax=26.604285…
 Xscl=5.5442857…
 Ymin=-9.32139
 Ymax=36.27
 Yscl=1
 Xres=1
```

Change Xmin to –18 and Xscl to 6 (sounds pretty reasonable).

```
WINDOW
 Xmin=-18
 Xmax=26.604285…
 Xscl=6
 Ymin=-9.32139
 Ymax=36.27
 Yscl=1
 Xres=1
```

To display the new graph, press [GRAPH]. NEVER press [ZOOM][9] after changing a window. You'll just go back to the one you had before! This looks better, but the bars at the right end are hard to see. Changing the scaling left some room at the top of the graph. Let's change the window and lower the Ymax to 25.

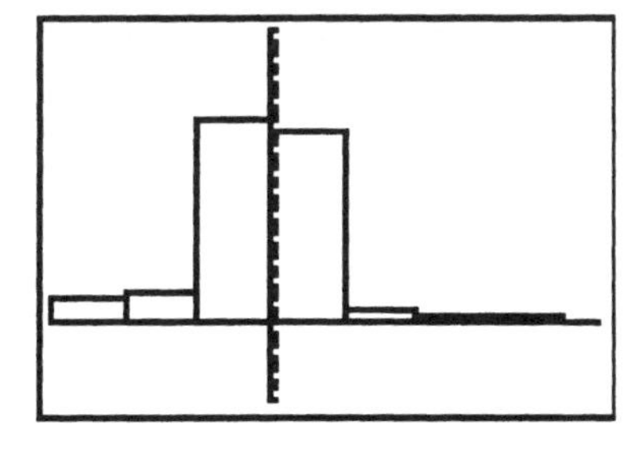

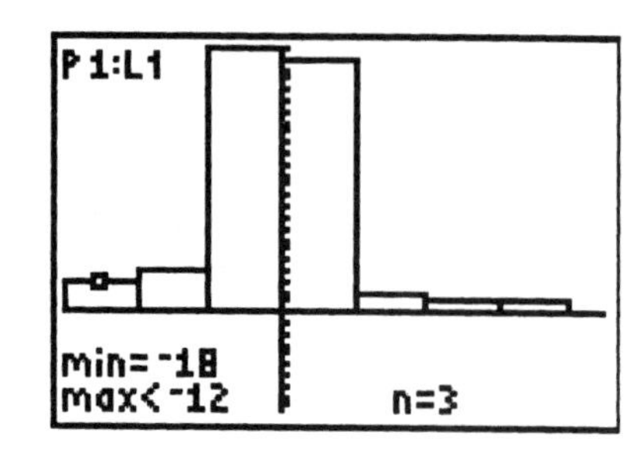

Here's the new graph (remember, press [GRAPH] after changing the window again). We still see the strong central peak, but now it looks like the right side of the graph is somewhat longer than the left. We can play with other settings, just to see how the shape might change. Notice that the gap at the right end has disappeared. Is there an outlier or not?

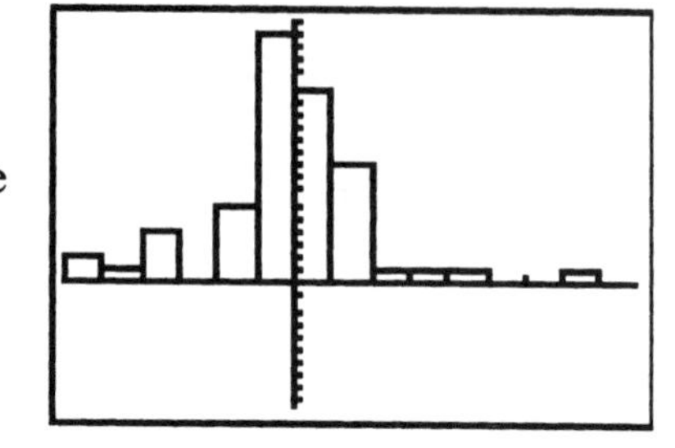

In this graph, Xscl (the barwidth) was set to 3. Ymax is 20.The distribution still looks unimodal, but somewhat right skewed. Notice the gap (two bars worth) on the high end. It looks like we have an outlier.

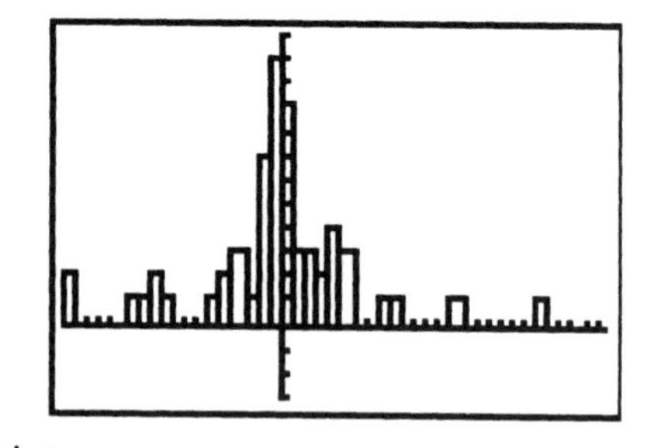

It's possible to have too many bars. Here, Xscl has been set to 1; Ymin is –3 and Ymax is 12. (These were changed for picture resolution.) We're starting to lose the forest for the trees.

How many bars and where are somewhat personal judgement. Your instructor may give you some guidelines. One "rule of thumb" for many years was to have somewhere between 5 and 20 bars; for most data sets dividing the number of observations by 5 gives a good estimate of how many bars will give a decent picture.

"Printing" the Picture

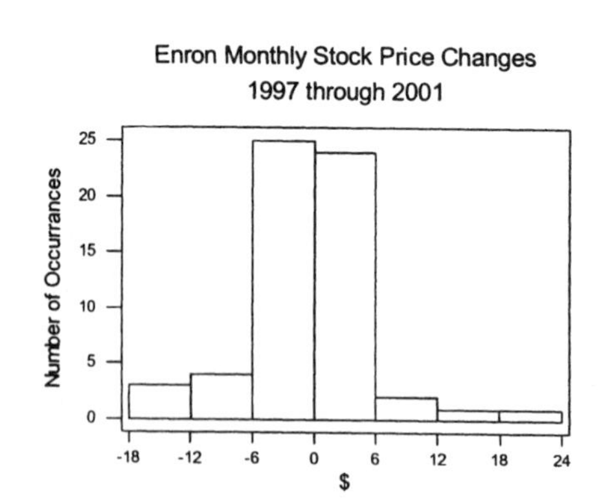

Unfortunately, calculators do not have printers. To make a hard copy of the graph once you are satisfied, use the [TRACE] key to examine the entire graph. Make a picture of the histogram, clearly labeling each axis and giving the graph a title. Remember that the intervals given are the endpoints of the intervals. Label them as such. When you are finished, you should have a picture like the one at right.

Histograms with Frequencies Specified

Sometimes data are given in the form of tables with both the data value and the number of times each value was observed. The frequency table below shows the heights (in inches) of 130 members of a choir. Entering 130 numbers could be tiresome, but there is a way to use the counts given.

We want to make a histogram to display this distribution. Enter the heights in one list and the observed counts in a second list. (We will put the heights in L1 and the counts in L2.) Make sure the lists are the same length, and that data values match with the given counts.

Height	Count
60	2
61	6
62	9
63	7
64	5
65	20
66	18
67	7
68	12
69	5
70	11
71	8
72	9
73	4
74	2
75	4
76	1

The first part of the lists looks like this.

L1	L2	L3 1
60	2	------
61	6	
62	9	
63	7	
64	5	
65	20	
66	18	

L1(1)=60

From [2nd][Y=] (Stat Plot) we will define Plot1 as at right. Notice Xlist is L1 (where the actual values are) and Freq is L2 (where the counts are.)

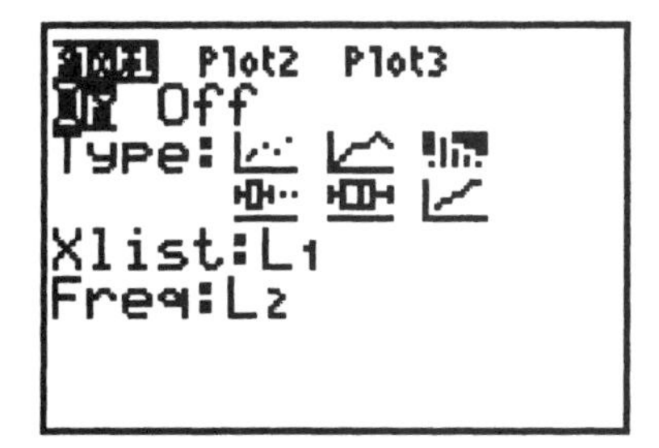

Pressing [ZOOM][9] gives the following graph. Notice that in this case the intervals and bar width (2) seem reasonable.

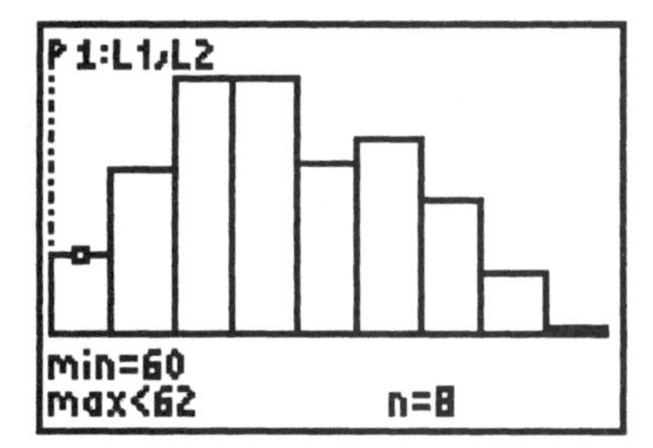

Histograms to Compare Distributions

We'd like to compare the distributions of two historically great baseball hitters: Babe Ruth and Mark McGwire. We have the following information on the numbers of home runs hit by Babe Ruth for 1920 through 1934 and for McGwire from 1986 to 2001.

Ruth: 52 59 35 41 46 25 47 60 54 46 49 46 41 34 22

McGwire: 3 49 32 33 39 22 42 9 9 39 52 34 24 70 65 32 29

[ZOOM][9] histograms of both distributions are below.

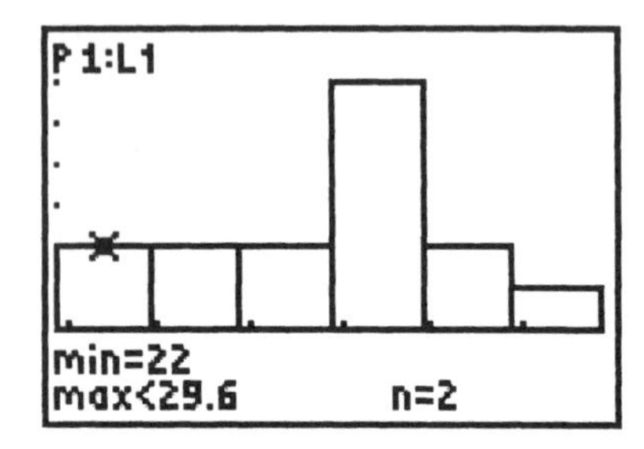

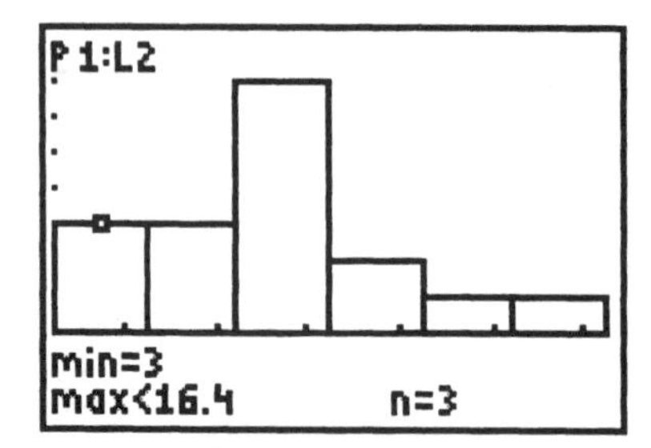

From the plots we can see both distributions are unimodal; Ruth's is skewed left and McGwire's is skewed right. But that's about all we can tell, because the graphs don't use the same scaling. People's eyes want to make a visual comparison, and since the graphs don't use the same values, this is impossible.

Let's change the scaling. Press [WINDOW]. Both the smallest and largest values occur in McGwire's distribution: 3 and 70. We also need a reasonable number of bars. It seems reasonable to set Xmin to 0, Xmax to 75 and use a bar width of 15. The settings used are at right. The rescaled plots are below.

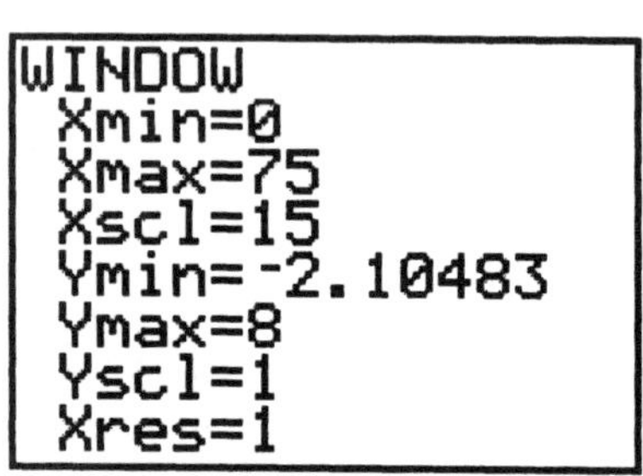

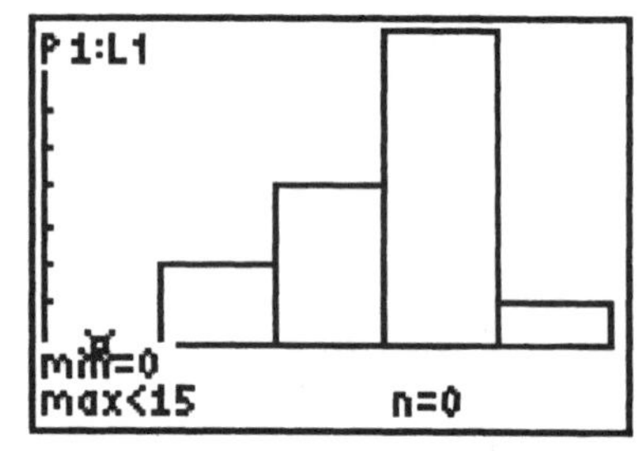

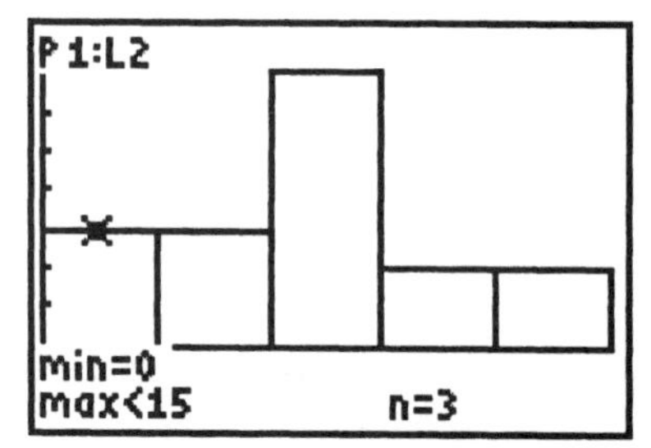

From these graphs it is easy to see that Ruth was the more prolific hitter. While McGwire had four years in which he hit more than 45 home runs, Ruth had 9. McGwire also had three seasons with fewer than 15 homers (due either to injury or his first, incomplete season).

Time Plots

Many variables are often measured at different points in time (as was the Enron data). It's not enough just to picture the distribution in this case. Time is an important factor, and we will want to know what (if any) part it plays. To answer this question, we will do a time (series) plot of the data. By convention, these are connected scatter plots with time represented on the x-axis and the actual variable values on the y-axis. They are connected because this makes any pattern easier to see than if the data points were just shown by themselves.

In our case, the time variable really consists of months and years, but TI calculators cannot handle this type of data. It will suffice for our purposes to simply use an index of months, from 1 to 60 to represent the five years' data. The data are in L1 and the time index has been entered into L2.

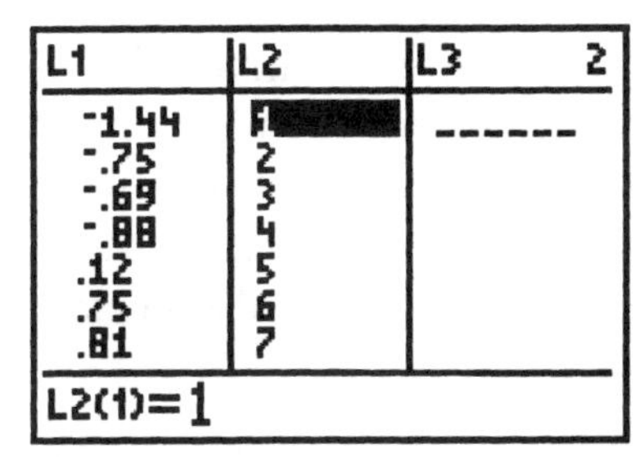

Connected scatter plots are the second plot type on the plot definition screen. The Xlist is the time indices and Ylist is the Enron values. There is a choice of three options for marking the actual data points. You may pick whichever one you like; however, from past experience we don't recommend the single pixel ever as it is too hard to distinguish. Once the plot has been defined, it can be displayed by pressing [ZOOM][9].

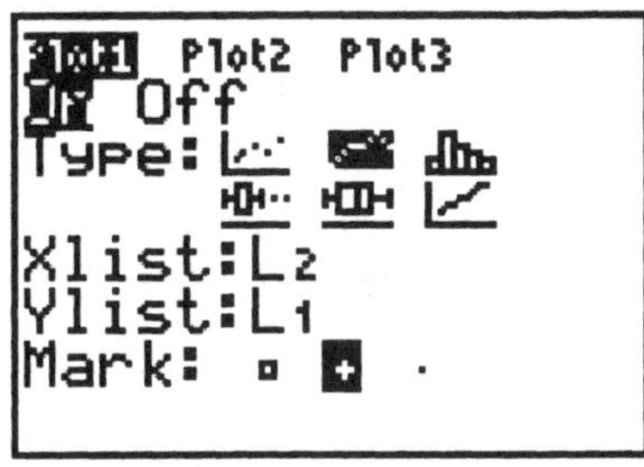

Here is the Enron Time Plot. We can clearly see that although there were fluctuations in price throughout the five-year period, these became much more dramatic (volatile) toward the end. In addition, it is clear that the price suffered steep declines toward the end.

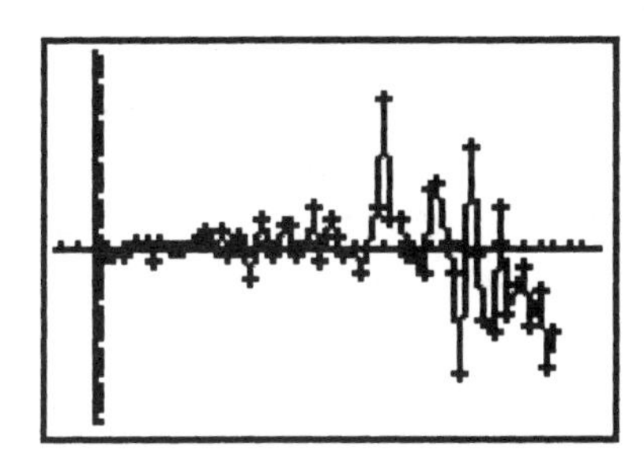

What can go wrong?

Help! I can't see the picture!

Seeing something like this (or a blank screen) is an indication of a windowing problem. This is usually caused by pressing [GRAPH] using an old setting. Try pressing [ZOOM][9] to display the graph with the current data.

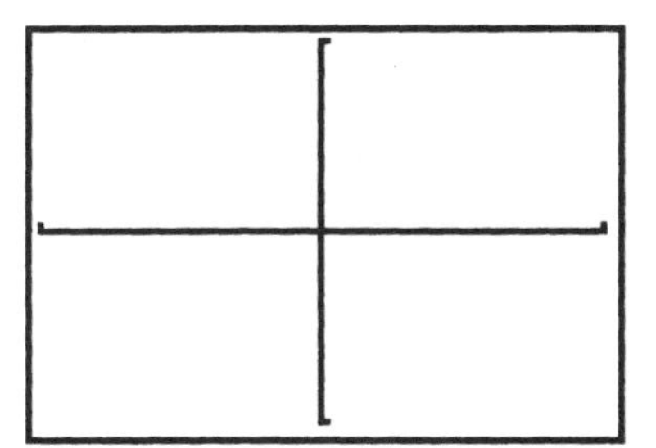

What's that weird line (or curve)?

There was a function entered on the [Y=] screen. The calculator graphs everything it possibly can at once. To eliminate the line, press [Y=]. For each function on the screen, press [CLEAR] to erase it. Then redraw the desired graph by pressing [GRAPH].

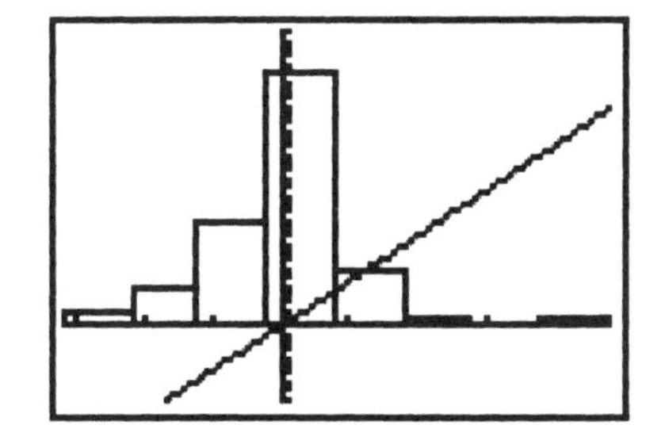

What's a Dim Mismatch?

This common error results from having two lists of unequal length. Here, it pertains either to a histogram with frequencies specified or a time plot. Press [ENTER] to clear the message, then return to the statistics editor and fix the problem.

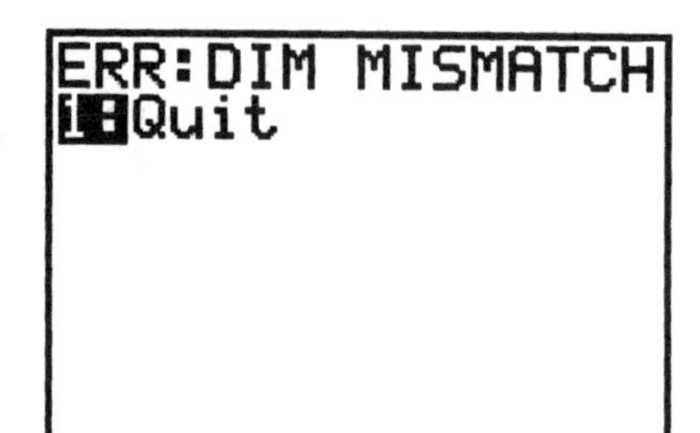

What's an Invalid Dim?

This problem is generally caused by reference to an empty list. Check the statistics editor for the lists you intended to use, then go back to the plot definition screen and correct them.

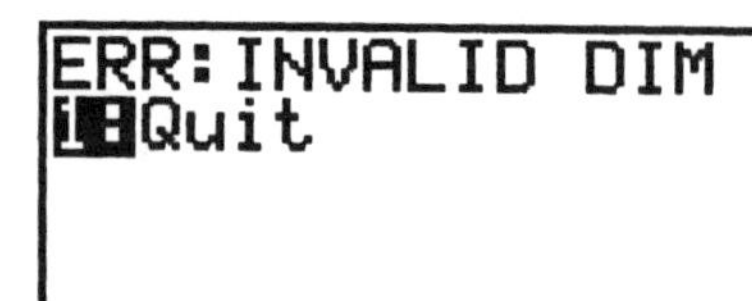

Stat?

This error is caused by having two stat plots turned on at the same time. What happened is the calculator tried to graph both, but the scalings are incompatible. Go to the STAT PLOT menu and turn off any undesired plot.

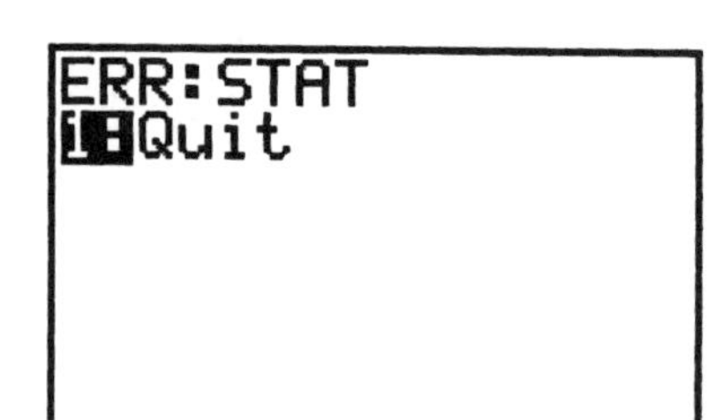

Chapter 3 – Describing Distributions Numerically

Remember the Enron data from the last chapter – monthly stock price changes for the five years preceding the crash. They are reproduced here for convenience.

	Jan	Feb	Mar	Apr	May	Jun	Jul	Aug	Sep	Oct	Nov	Dec
1997	-$1.44	-0.75	-0.69	-0.88	0.12	0.75	0.81	-1.75	0.69	-0.22	-0.16	0.34
1998	0.78	0.62	2.44	-0.28	2.22	-0.50	2.06	-0.88	-4.50	4.12	1.16	-0.50
1999	3.28	3.34	-1.22	0.47	5.62	-1.59	4.31	1.47	-0.72	-0.38	-3.25	0.03
2000	5.72	21.06	4.50	4.56	-1.25	-1.19	-3.12	8.00	9.31	1.12	-3.19	-17.75
2001	14.38	-1.08	-10.11	-12.11	5.84	-9.37	-4.74	-2.69	-10.61	-5.85	-17.16	-11.59

What is a typical "middle" value? What is the spread of the data? These are the questions addressed in this chapter. We will also meet a new statistics plot based on numerical summaries.

Calculating Numerical Summaries

To calculate the numerical summary statistics for a single variable, first enter them into a list.
Here the data have been entered into list L1. The first few values are shown at right.

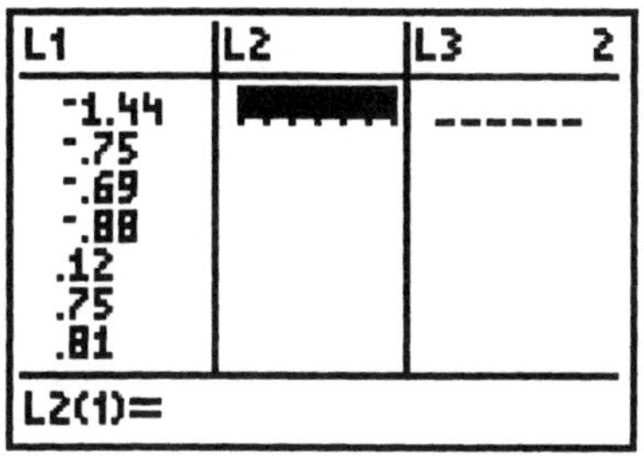

Press STAT then arrow to CALC. The menu at right will be displayed. The menu is organized so that the most often used options are at the top. Notice that 1:1-Var Stats is highlighted. Press ENTER to select that option (or press 1). The command will be transferred to the home screen.

Now you need to tell the calculator which list is to be used as input. Press 2nd 1 (L1). If no list name is given, the calculator will default to use L1, but it's good practice to get into the habit of specifying the list name.
Press ENTER to carry out the command.

The first page of results is displayed at right. The arrow at the bottom left indicates more results are available and can be found by using ▼. We first see the mean monthly stock price change is -$0.3733333. The calculator does not know if the data you are using represent a sample or a population. It has only one symbol for the mean ($\bar{x}$). If your data represents a population, you should report the mean using proper notation, and call it μ. The two values displayed next are the sum of the data values and the sum of squared data values. These are intermediate quantities used in computations, and are generally not of interest. Two different standard deviations are also reported as measures of spread.

```
1-Var Stats
 x̄=-.3733333333
 Σx=-22.4
 Σx²=2345.7652
 Sx=6.294203054
 σx=6.241530973
↓n=60
```

$Sx = \sqrt{\dfrac{\sum (x_i - \bar{x})^2}{(n-1)}}$ is the sample standard deviation and σx is the population standard deviation (the formula is the same, except the divisor is n). This data does not represent all possible monthly stock changes for Enron, so we will use s_x (6.2942) as the standard deviation. Which is the correct value to use depends on data you have – is it from a sample or is it for a population? The last value on the screen is the number of items in the data list, n, is 60.

One thing to bear in mind is that calculators (and computers) will use (and report) many more digits than really make sense to use. It comes from division (in which, as we know, things don't always come out evenly) and taking square roots (which generally aren't whole numbers). How many digits to report should be decided by your instructor, but a good rule of thumb is to report one more place than in the original data. Our data was in dollars and cents, so we'll use three decimal places. Also, since we don't have all the possible monthly stock price changes for Enron, we will report $\bar{x}$ = $-0.373 and s = $6.294.

Using the down arrow, we find the five-number summary. The median (another measure of center) is -$0.25, which is close to the mean in this data set (as it should be since the data were roughly symmetric). We can use the other values in this summary to compute two other measures of spread: the Interquartile Range (IQR) which is the spread of the middle half of the data, and the Range. The IQR is $Q_3 - Q_1$, or $2.14 – (-$1.67) which is $3.81. This means the central half of the data had a spread of $3.81. The Range is max – min, or $21.06 – (-$17.75) which is $38.81.

```
1-Var Stats
↑n=60
 minX=-17.75
 Q1=-1.67
 Med=-.25
 Q3=2.14
 maxX=21.06
```

Statistics for Tabulated data

In the last chapter we looked at the distribution of heights of members in a choir. They were presented in a table of heights along with how many choir members there were of a given height. With these data in lists L2 and L3, we would like to know the average height for the choir.

Just as before, press STAT then arrow to CALC, press ENTER to select choice 1:1-Var Stats.
On the home screen, you will specify not just one list, but two. The first list is the list of values (L2) and the second is the list of counts (L3). Your command will look like the screen at right. Don't worry that it didn't all fit on a single line. Press ENTER to execute the command.

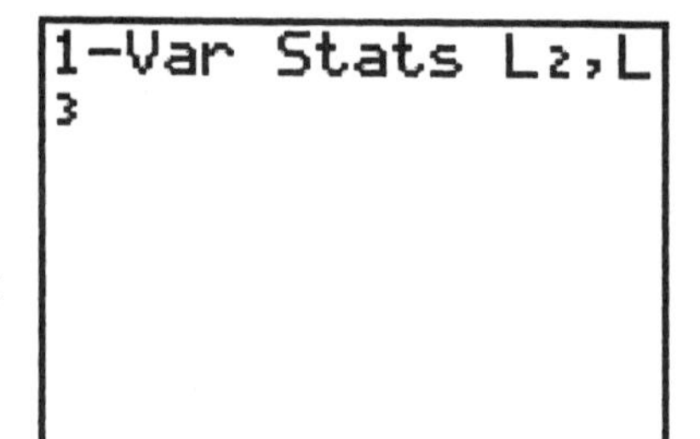

Here are the results. The average (mean) height for the choir members was 67.1 inches. The standard deviation (assuming these are not all the members possible for the choir) is 3.8 inches. Paging down, we find the median height was 66 inches. It is not surprising the median would be somewhat less than the mean for these data since the histogram indicated a right skewed distribution.

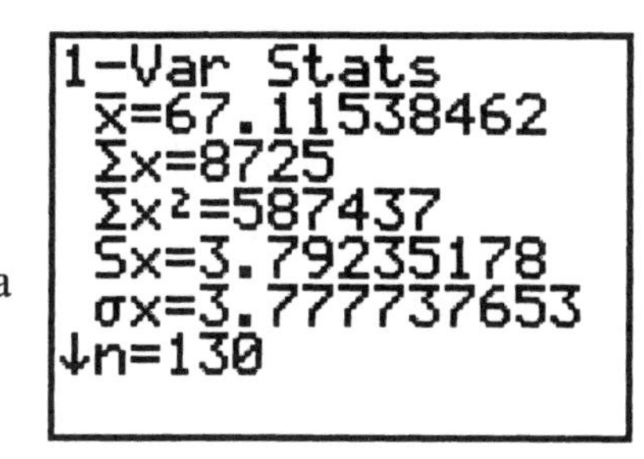

Boxplots

Box plots (sometimes called box-and-whisker plots) are another way of picturing a distribution. Unlike histograms, they are based on definite values and are not subjective. However, as no plot is perfect, they can hide some potential features such as bimodality. A good practice, since it is generally so easy, is to look at several plots. They all can show different things.

There are two types of boxplots – the original which is based on the five-number summary (min, Q_1, median, Q_3, and max) and a "modified" boxplot which has an objective criterion to identify outliers. Both types of plots divide the data into fourths – a "whisker" for the bottom and top quarters of the data, and a box for the middle half, with the median indicated inside the box. We always recommend using the modified boxplot, but your instructor may suggest otherwise.

As always, begin with data in a list. We will begin with the Enron data already examined. The data are in L1. Press [2nd][Y=] (StatPlots) then [ENTER] to get to the plot definitions screen for Plot 1.

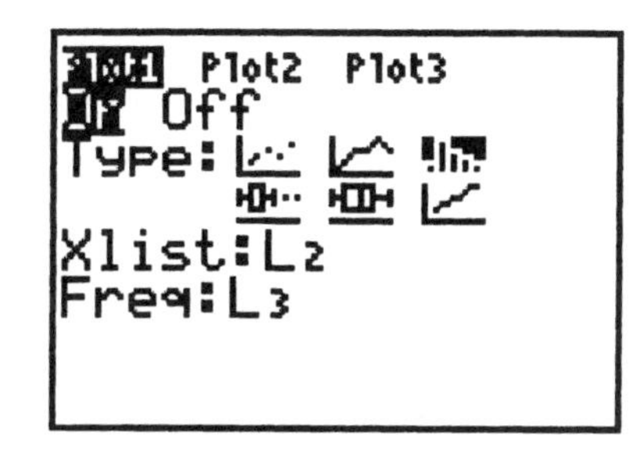

Notice there are two choices for boxplots - which is the boxplot where outliers are identified and which does not identify outliers. We will look at both to see the difference between the two.

Move the cursor to highlight . Make sure Xlist is changed to L1 (press [2nd][1]); also make sure Freq: is set to 1 (press [ALPHA][1] if necessary). Your plot definition screen should look like the screen at right.

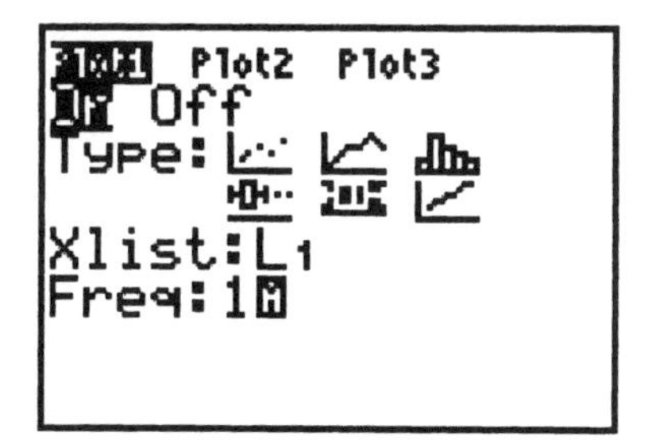

Press [ZOOM][9] to display the graph. Here we see a short central box, indicating the middle half of the data is close to 0, and long relatively even whiskers. The median (-0.25) is very close to zero and is hard to distinguish. The indications from the plot are that the distribution is symmetric. Pressing [TRACE] and using the right and left arrows will allow you to move around the graph, locating the median, quartiles, min, and max.

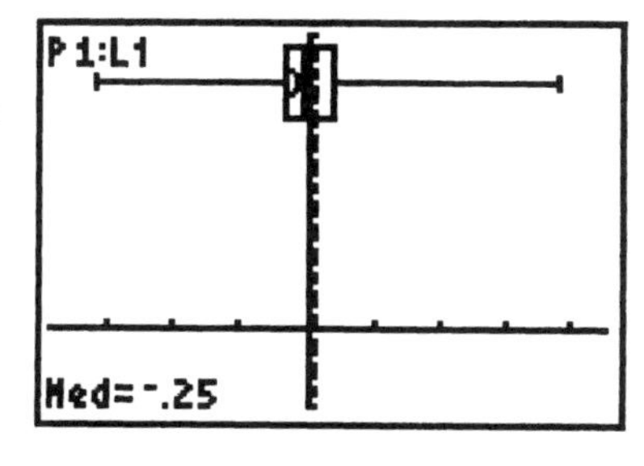

Return to the Plot definition screen and change the plot type to a boxplot identifying outliers (·□···). Pressing [ZOOM][9] will display the plot at right. The central box has not changed; the whickers are shorted and many points on either end have been flagged as outliers by the $Q_3 + 1.5*IQR$ and $Q_1 - 1.5*IQR$ criteria.

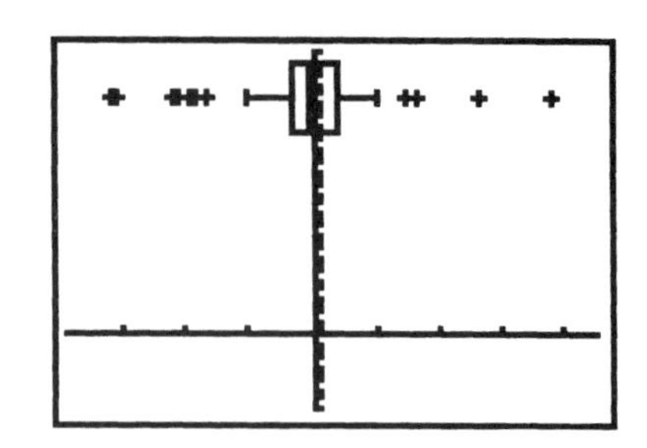

Boxplots with Tabulated Data

Reconsider the data on heights of members of a choir. According to the histogram, this was somewhat right-skewed. What will its boxplot look like? With heights in L2 and frequencies in L3, the plot definition screen looks like that at right.

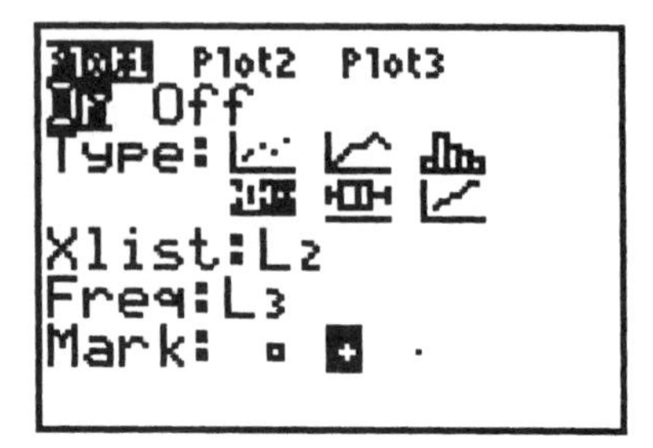

After pressing [ZOOM][9] the graph at right will be displayed. Notice the median is not in the middle of the box; the right half of the data is longer than the left half (the data is right skewed) even though the two whiskers are relatively equal in length. Also, since we defined the plot to identify any possible outliers, none are flagged, so these data have no outliers.

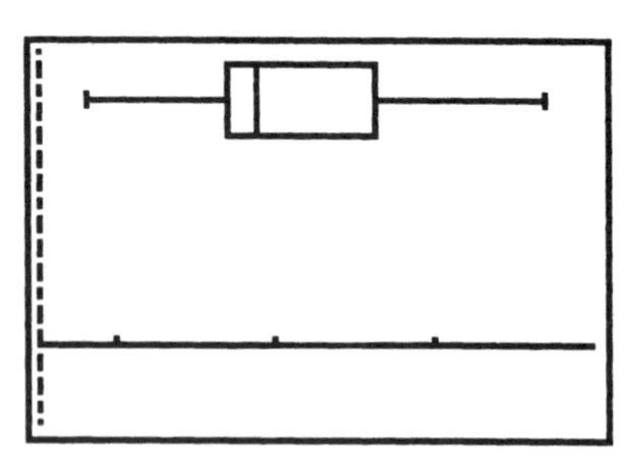

Boxplots to Compare Distributions

Boxplots are very useful in comparing distributions. This is one of the few exceptions to the rule about only one plot being turned on at a time. Up to three boxplots can be displayed at once. Displaying boxplots side-by-side is a good visual comparison of distributions.

Recall the data for Babe Ruth and Mark McGwire. How would these distributions look displayed together as boxplots? The data have been entered into L4 (Ruth) and L5 (McGwire). We define Plot 1 to use Ruth's home run values as at right.

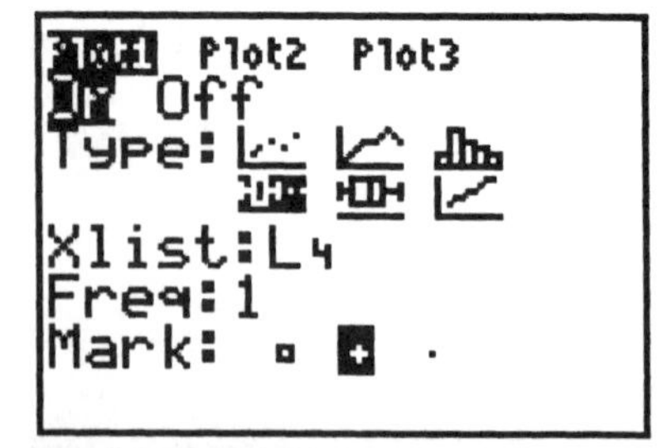

Returning to the STAT PLOT menu, we will arrow down to Plot 2, and define it to use McGwire's numbers as at right.

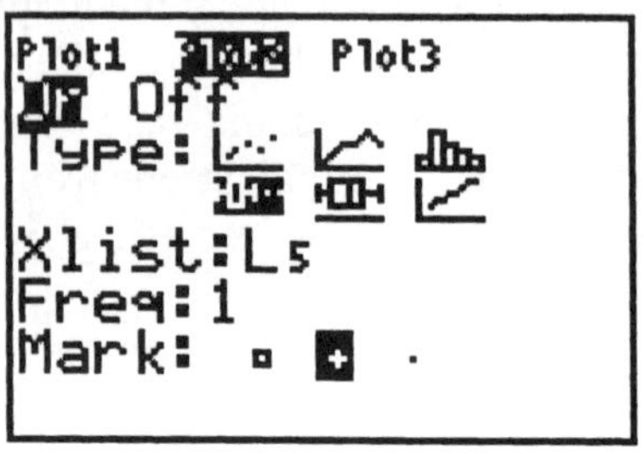

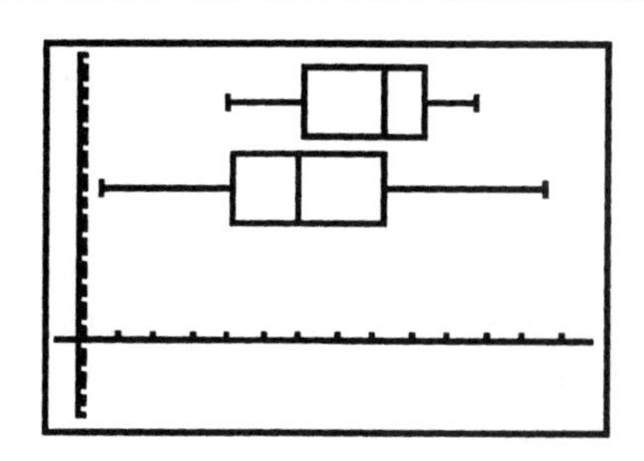

Pressing [ZOOM][9] gives the display at right. The top plot is Ruth's home run distribution; the bottom is McGwire's. (Pressing [TRACE] will identify each plot; to move from one to the other, press the down and up arrows.) Neither distribution has outliers. Ruth's is much less variable than McGwire's and is skewed left. McGwire's distribution appears rather symmetric.

Chapter 4 – The Standard Deviation as a Ruler and the Normal Model

The standard deviation is the most common measure of variation; it plays a crucial role in how we look at data. *Z*-scores measure standard deviations above or below the mean and are useful as measures of relative standing. Normal models are very useful as many random variables (at least approximately) follow its unimodal, symmetric shape.

Z-Scores

The *z*-score for an observation is $z = \frac{(obs - \bar{y})}{s}$, where *obs* is the value of interest. Positive values indicate the observation is above the mean; negatives mean the value is below the mean. Calculating them is easy as long as one keeps in mind that the subtraction in the numerator must be done before the division. Calculators follow the arithmetic hierarchy of operations.

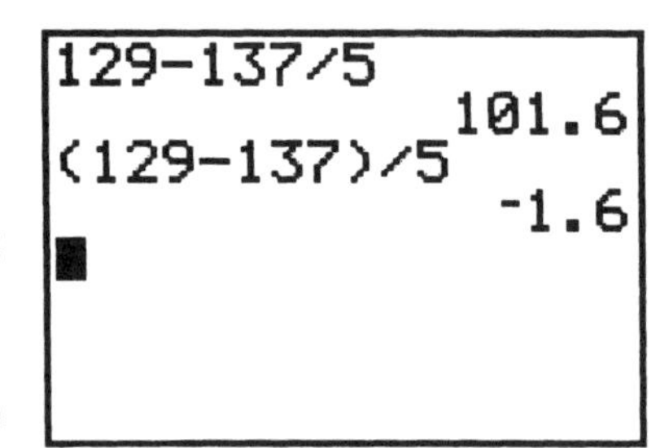

For example, in winning an 800m race during the Olympic women's heptathlon, Gertrud Bacher of Italy had a time of 129 seconds; the mean time for all runners in the race was 137 seconds with standard deviation 5 seconds. What is Bacher's *z*-score? Two examples of the calculation are at right. The first (incorrect!) indicates she was 101.6 standard deviations *above* the mean – unreasonable since she won the race, so had the fastest time. The problem is failing to perform the subtraction first or enclosing the numerator in parentheses. The second (correct!) calculation indicates Bacher's time was 1.6 standard deviations *below* the mean.

Working with Normal Curves

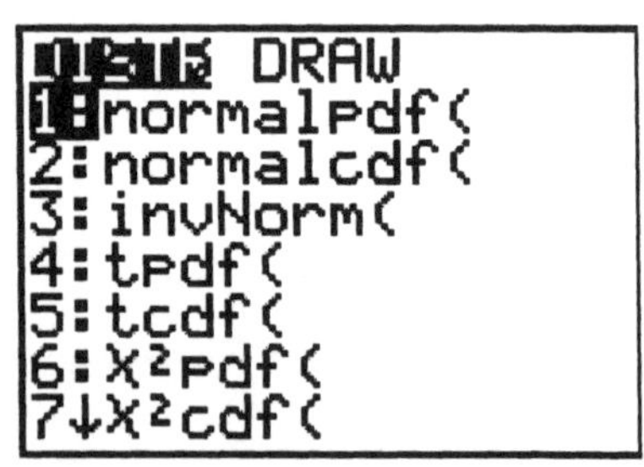

What proportion of SAT scores are between 450 and 600? SAT scores for each of the two tests (verbal and math) are approximately normal with mean 500 and standard deviation 100, or N(500, 100). There are two ways to answer this question with a TI-83 calculator. One will draw the curve; the other just answers the question. Both start at the same place: the Distributions menu. Press [2nd][VARS] (DISTR) The screen at right will appear. Notice the arrow pointing down at the bottom left. There are more distributions which can be used; more will be said about some of them later. At the top of the screen, there are two choices DISTR (the default) which merely gives distribution values, and DRAW which will draw the curves and shade the appropriate areas.

First, let's just answer the question. The menu option to select is 2:normalcdf(. Either press the down arrow and then [ENTER] or just press [2]. The command will be transferred to the home screen. It normally requires two parameters to be entered: the z-score for the low end of the area of interest and the z-score for high end. Separate the entries by commas, and finish by closing the parentheses.

A score of 450 is 0.5 standard deviations below the mean, so its z-score is –0.5. A score of 600 is 1 standard deviation above the mean; its z-score is 1.

```
normalcdf(-.5,1)
          .5328072082
```

The command has been entered in the screen at right, [ENTER] was pressed to execute the command. We see that about 53.3% of all scores on the SAT will be between 450 and 600.

To find the area and have it shaded, one needs to first set the window ([ZOOM][9] does not work here). For any normal model, values will range from about –3 to 3 (three standard deviations either side of the mean). Since the whole area under the curve is 1, the height of the curve will be a small number; we have set the Ymin to –0.1 and Ymax to 0.4 as in the screen at right.

```
WINDOW
 Xmin=-3
 Xmax=3
 Xscl=1
 Ymin=-.1
 Ymax=.4
 Yscl=.1
 Xres=3
```

Now press [2nd][VARS] (DISTR) and arrow to DRAW. We want choice 1, so press [ENTER].

```
DISTR DRAW
1:ShadeNorm(
2:Shade_t(
3:ShadeX²(
4:ShadeF(
```

The command has been transferred to the home screen. Enter the z-scores for the low end of interest and the high end as at right, then press [ENTER].

```
ShadeNorm(-.5,1)
```

The graph should look like the one at right. Again, we see the area is about 53.3%; we also see what portion of the normal curve it represents.

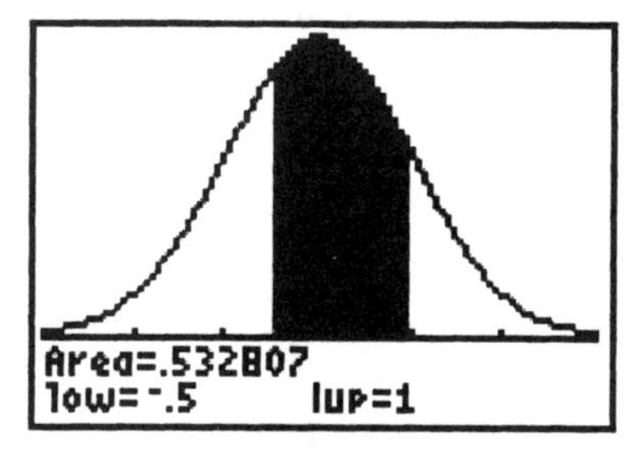

When working with the DRAW option, the graph must be cleared between successive commands or the shaded area will accumulate until the whole curve is shaded. To clear the drawing, press [2nd][PRGM] (DRAW) then press [ENTER] to select option 1:ClrDraw.

Let's look at another example. A cereal manufacturer makes boxes labeled as 16 ounces; but the boxes are actually filled according to a normal model with mean 16.3 ounces and standard deviation 0.2 ounces. We want to know what fraction of all boxes will be "underweight," that is, contain less than the advertised 16 ounces.

Strictly speaking, Normal models extend from $-\infty$ to ∞ (negative infinity to infinity). On the calculator, ∞ is represented as 1e99 (10^{99}). To enter this, one presses [1][2nd][,][9][9], but practically, any "very large" negative number (say, -99) will work for $-\infty$ and any large positive number (say 99) for ∞ since we know almost all of the area is between –3 and 3.

We want to know what fraction of all boxes are less than 16 ounces, so the low end of interest is $-\infty$ (we entered –99 as the stand-in); the upper end of interest is 16 which corresponds to a z-score of –1.5. The command and the result are at right. We see that about 6.7% of all boxes of this cereal should be underweight.

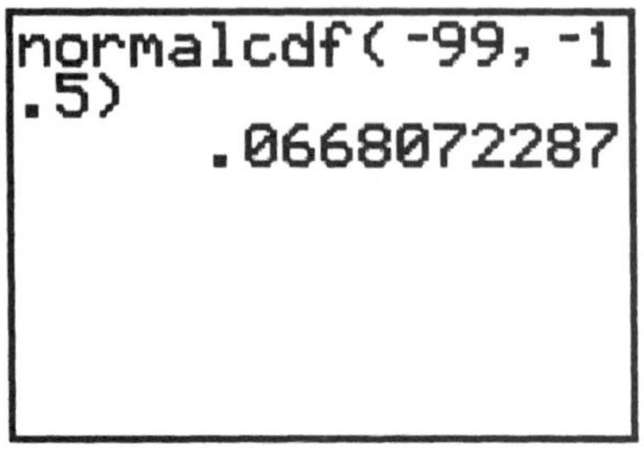

```
normalcdf(-99,-1
.5)
          .0668072287
```

Working with Normal Percentiles

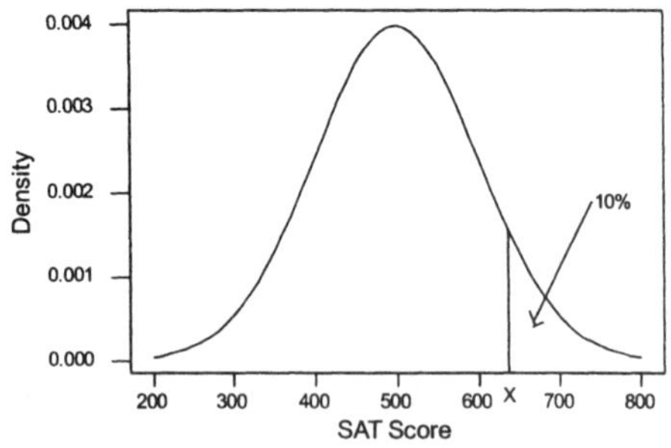

Sometimes the area under the curve is given and the corresponding value of the variable is of interest. For example, in the SAT model used before, how high must a student score to place in the top 10%? In a sketch of the normal curve, the unknown value, we'll call it X, separates the top 10% from the lower 90%. We first have to find a corresponding z-score.

This is the inverse situation from that we've just explored. On the DISTR menu, the command is `3:invNorm(`. Press [2nd][VARS] then [3] to transfer the command to the home screen. The parameter for this command is area to the left of the point of interest (.90 or 90%). Press [ENTER] to execute the command. The z-score of interest is 1.28. To be in the top 10%, your score must be 1.28 standard deviations above the mean.

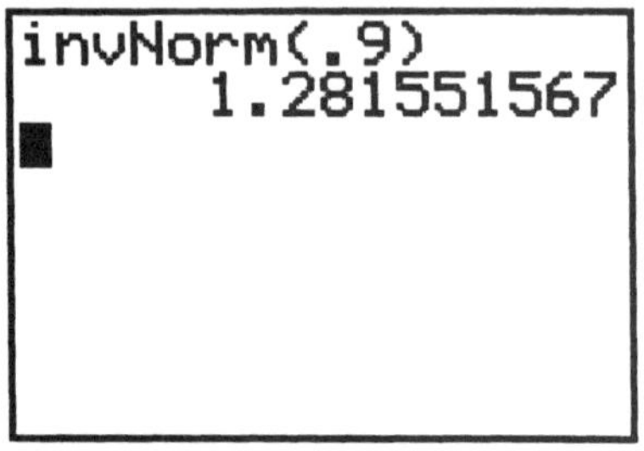

```
invNorm(.9)
          1.281551567
■
```

$z = (x - \mu)/\sigma = 1.28$

$(x - 500)/100 = 1.28$

After doing the algebra, we see that a score of 628 will put a person in the top 10% of all SAT scores; practically since scores are reported rounded to multiples of 10, a score of 630 is needed.

Here is another example. The cereal company's lawyers are not happy with 6.7% of boxes being underweight. They want at most 4% to be underweight. What mean must the company reset its machines to in order to achieve this target? We'll use `invNorm` for a standard normal model to find the z-score corresponding to 4% of the area below this value, then use some algebra to solve for the unknown mean.

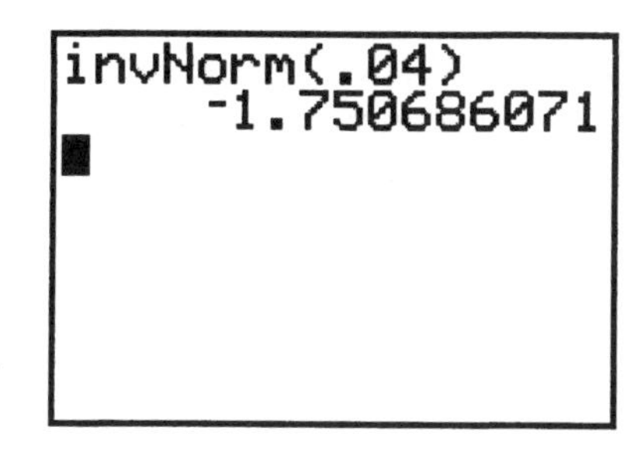

The z-score of interest is −1.75.

$z = (x - \mu)/\sigma = -1.75$

$(16 - \mu)/0.2 = -1.75$

Now multiplying both sides by 0.2, and subtracting 16 from both sides gives

$-\mu = -16.35$ or $\mu = 16.35$. In order to achieve the target of no more than 4% of boxes being underweight, the machine will have to be set for an average of 16.35 ounces per box.

Is my data Normal?

It is one thing to assume data follows a normal model. When one actually has data this should be checked. One method is to look at a histogram – is it unimodal, symmetric and bell-shaped? Another is to ask whether the data (roughly) follow the 68-95-99.7 rule. Both of these might work well with a fairly large data set; however, there is a specialized tool called a normal probability plot that will work with any size data set. This plots the data on one axis against the z-score one would expect if the data were exactly normal on the other. If the data is normal this plot will look like a diagonal straight line.

Recall the data on Enron stock price changes. They are in list L1. Press [2nd][Y=] to get to the first Stat Plot screen. Once here, you should always check that all plots are off except the one you will use. Select the plot to use and press [ENTER]. The normal probability plot is the last plot type. Use the right arrow to move there and press [ENTER] to move the highlight. Notice you have a choice of having the data on either the X or Y axis. It doesn't really matter which you choose; many statistical packages put the data on the x-axis; many texts (including DeVeaux and Velleman) put the data on the y-axis. As we have seen before, you have a choice of three marks for each data point. Select the one you prefer.

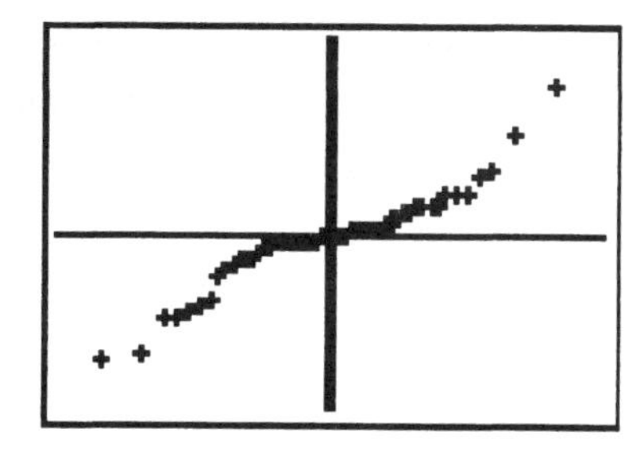

Pressing [ZOOM][9] displays the graph. This graph is not a perfect straight line, and shows some stragglers (outliers) at each end just as we've seen before. I would hesitate to call this data normal.

Skewed distributions often show a curved shape. Data on the cost per minute of phone calls as advertised by Net2Phone in USA Today (July 9 2001) to 22 countries were as follows:

7.9	17	3.9	9.9	15	9.9	7.9	7.9	7.9	7.9	8.9
21	6.9	11	9.9	9.9	7.9	3.9	22	9.9	7.9	16

We have entered in a list and have defined the normal probability plot as above.

The plot obtained is at right. Not only does it show a general upward curve, it also displays something called *granularity*. This occurs when a particular data value occurs several times (as with 7.9 cents per minute which was in the list 7 times.)

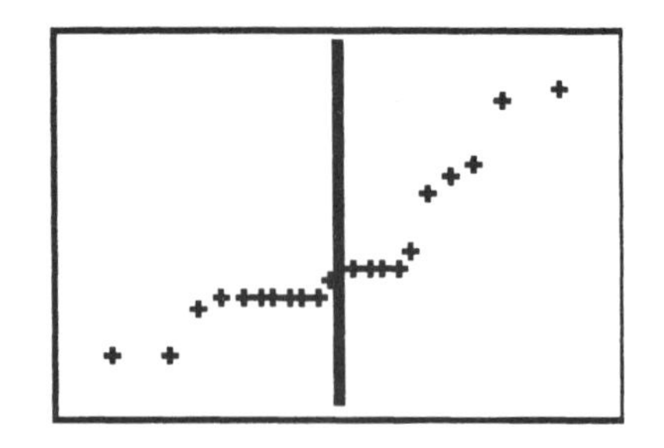

What can go wrong?

Why is my curve all black?

For the SAT scores curve, the graph indicates more than half of the area is of interest between 200 and 475 (z-scores of –3 and -.25); the message at the bottom says the area is 40%. This is a result of having failed to clear the drawing between commands. Press [2nd][PRGM] then [ENTER] to clear the drawing, then reexecute the command.

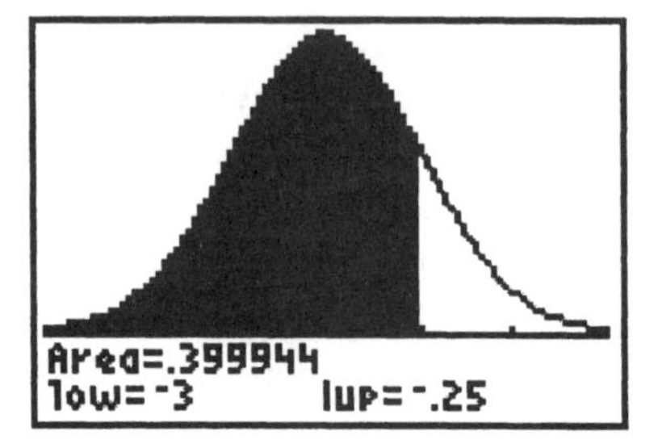

How can the probability be more than 1?

It can't. If the results look like the probability is more than one, check the right side of the result for an exponent. Here it is –4. That means the leading 2 is really in the fourth decimal place, so the probability is 0.0002. The chance a variable is more than 3.5 standard deviations above the mean (this would be a box of the cereal more than 17 ounces) is about 0.02%.

```
normalcdf(3.5,99
)
      2.326733735E-4
```

How can the probability be negative?

It can't. The low and high ends of the area of interest have been entered in the wrong order. As the calculator does a numerical integration to find the answer, it doesn't care. You should.

```
normalcdf(1,-99)
        -.8413447404
```

What's Err: Domain?

This message comes as a result of having entered the `invNorm` command with parameter 90. (You wanted to find the value that puts you into the top 10% of SAT scores.) The percentage must be entered as a decimal number. Reenter the command with parameter .90.

Chapter 5 – Scatterplots, Correlation, and Regression

Are two numeric variables related? If so, how? Scatterplots and regression will answer these questions. Correlation describes the direction and strength of linear relationships. Linear regression further describes these relationships.

Here are advertised horsepower ratings and expected gas mileage for several 2001 vehicles.

Audi A4	170 hp	22 mpg	Buick LeSabre	205	20
Chevy Blazer	190	15	Chevy Prism	125	31
Ford Excursion	310	10	GMC Yukon	285	13
Honda Civic	127	29	Hyundai Elantra	140	25
Lexus 300	215	21	Lincoln LS	210	23
Mazda MPV	170	18	Olds Alero	140	23
Toyota Camry	194	21	VW Beetle	115	29

How is horsepower related to gas mileage? The first step in examining relationships is through a scatterplot.

Scatterplots

Here are the first few values for the horsepower ratings (in L1) and the gas mileage (in L2). It is important to enter these very carefully as they have been entered in the table, since the values represent a data pair for each vehicle type.

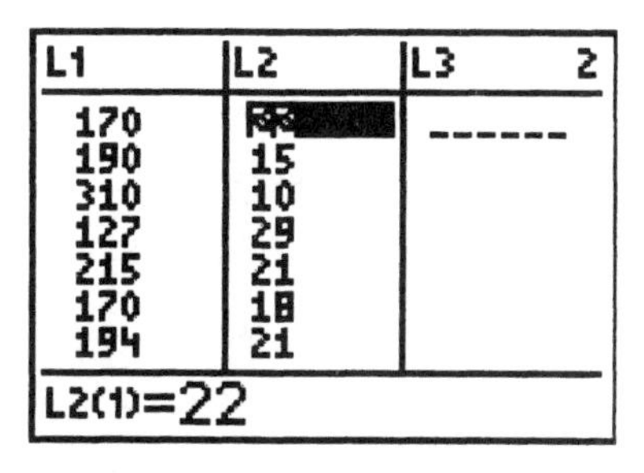

Our supposition is that larger engines will get less gas mileage, so we will use the horsepower ratings as the predictor (X) variable and the gas mileage as the response (Y) variable.

To define the scatterplot, press [2nd][Y=] (STAT PLOT). Select Plot1 by pressing [ENTER]. Scatterplots are the first plot type. Move the cursor to highlight that plot, and press [ENTER] to move the highlight. Press the down arrow ([▼]) and enter the list where the predictor (*X*) variable is (here, L1, so [2nd][1]). Press the down arrow and enter the list containing the response (*Y*) variable (here, L2, so [2nd][2]). Press [▼] to select the type of mark for each data point (*X, Y*) pair. The author recommends either the square or cross; the single pixel tends to be too hard to see. When finished, the plot definition screen should look like the one at right.

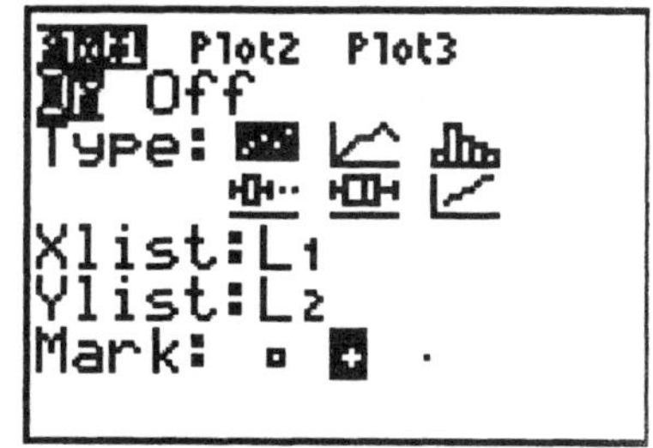

Press [ZOOM][9] to view the plot. Here we see a generally decreasing pattern from left to right, supporting our initial idea. The pattern is generally linear; however, two points at the bottom right may be unusual; we'll examine those later.

Regression Setup

We want to examine the nature of this relationship further; before we do, we need to set up the calculator to display the values of the correlation coefficient (r) and the coefficient of determination (r^2).

This procedure *normally* needs to be done only *once*; however, changing batteries slowly will reset memory and it may have to be done over.

Press [2nd][0] (Catalog). This accesses the list of all the functions the calculator knows about. Notice that the cursor is at the beginning of the catalog. We need to get down to a command that begins with a D, so press [x^{-1}] (alpha D).

```
CATALOG
▸abs(
 and
 angle(
 ANOVA(
 Ans
 augment(
 AxesOff
```

We're now here. We haven't gotten to the command yet, but we're close.

```
CATALOG
▸dbd(
 ▸Dec
 Degree
 DelVar
 DependAsk
 DependAuto
 det(
```

Press the down arrow [▼] until the command DiagnosticOn is highlighted. Press [ENTER] to select the command and transfer it to the home screen, then [ENTER] again to execute it.

```
CATALOG
 Degree
 DelVar
 DependAsk
 DependAuto
 det(
 DiagnosticOff
▸DiagnosticOn
```

Your screen should look like the one at right.

```
DiagnosticOn
            Done
```

Regression and Correlation

We're now ready to examine the correlation between these two variables. However, the calculator will not give just the value of r; it's much easier computationally for it to do the whole thing at once and report all the values of interest.

Press [STAT], arrow to CALC. (We've been here before for `1-var Stats`). There are two linear regression choices: 4:LinReg(ax+b) and 8:LinReg(a+bx). The answers you get will be the same, but one must keep in mind the order in which the coefficients are used. Since statisticians usually prefer the constant term of the regression to come first (in case there are several predictor variables – multiple regression) we'll use choice 8.

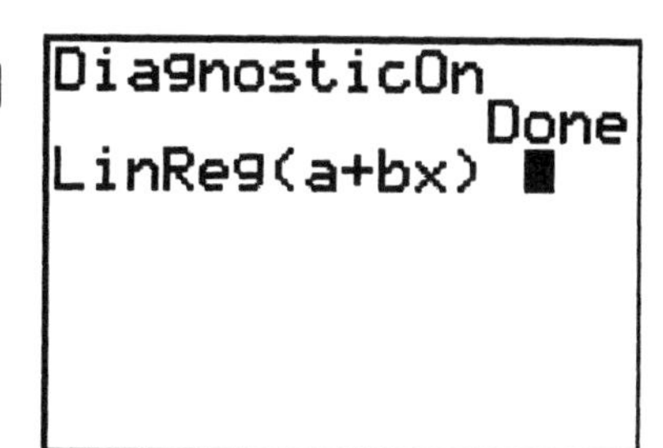

Either press the down arrow until the selection is highlighted then [ENTER] or simply press [8]. The command will be transferred to the home screen. In doing a simple regression with predictor variable in L1 and response in L2, simply pressing [ENTER] at this point is enough. However, if you want to store the equation (to see it on your graph for one reason), you need to specify the lists in which the data is stored; it's also just good practice to get into the habit, since you may want to use other lists than these.

```
DiagnosticOn
                Done
LinReg(a+bx) L1,
L2
```

Press [2nd][1] (L1), then [2nd][2] (L2) followed by [ENTER] to execute the command. Before pressing [ENTER] the screen should look like the one at right.

```
LinReg
 y=a+bx
 a=38.45416426
 b=-.0918175268
 r²=.7708040509
 r=-.8779544697
```

Once the command is executed, you should see the display at right. Notice the first line of the results displays the type of regression in terms of y and x. Regression lines should never be reported in these terms, but the calculator does not know what variables you are working with. This is really an aid to remind you where the coefficients a and b go in the equation.

Here, we have the following regression equation:
$Mileage = 38.45 - 0.09 * horsepower$. Remember, this line represents the average value of gas mileage for a given horsepower rating, based on the model from our data. In addition, we see the correlation coefficient, r = -0.878 which indicates a strong, negative relationship. The coefficient of determination, r^2 = 0.771 (normally expressed as r^2 = 77.2%) tells us that 77.2% of the observed variation in gas mileage (remember these values range from 10 to 31 mpg) is explained by the model.

Storing the regression line

It was previously mentioned that the equation of the regression line can be stored for future reference. This is done by modifying the regression command as follows:

Press [STAT], arrow to CALC, then [8][2nd][1][,][2nd][2] (so far, this is what we did before). Now press [,][VARS] arrow to Y-Vars, press [ENTER] to select 1:Function, then [ENTER] to select Y_1. The regression command should look like that at right. Press [ENTER] to execute the command.

The regression output will look the same as before. The difference between the two commands can be seen by pressing [Y=]. The equation of the line has been stored for further use.

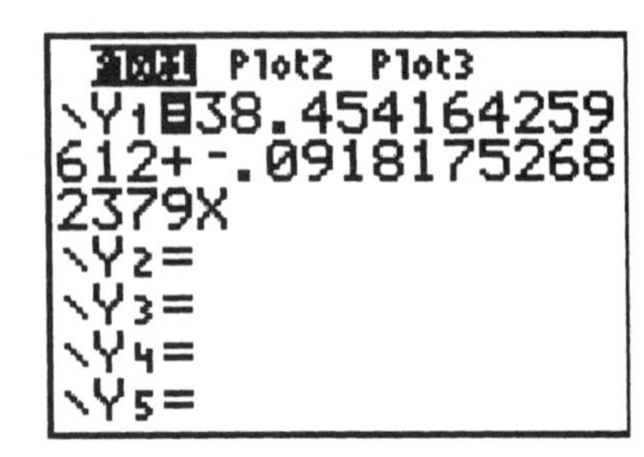

It would be nice to see how the line passes through the data; it should be roughly in the center of the data points. Press [GRAPH], since there is no need to resize the window. Sure enough, there's the line just as we expected. Notice that since no line will be perfect (unless r = ±1), some of the points are above the line, and some below. The distances between the points and the regression line are called residuals and their plots are used to examine the line for adequacy of the model.

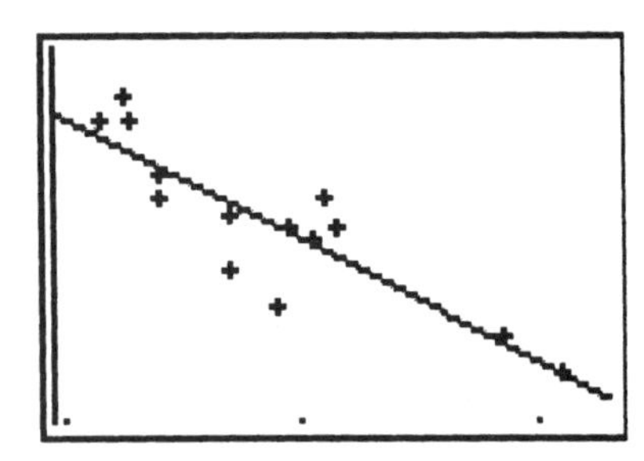

Residuals Plots

Residuals are defined to be the vertical distance from the data point to the regression line, in other words, $e_i = y_i - (a + bx_i)$ for each data point (x_i, y_i) in the data set. The e_i are the residuals. There are two ways to obtain the residuals. The first makes use of the stored regression line. From the list of y-values we will subtract the value on the line by "plugging in" each corresponding x-value into the equation. The residuals will then be stored into a new list. For our example we will enter the following command onto the home screen: [2nd][2] [−] [VARS], arrow to Y-Vars, press [ENTER]to select Function, then [ENTER] to select Y_1[(][2nd][1][)][STO▸][2nd][3]. This command says "take the y-list and from it subtract the value obtained by evaluating function Y_1 at each x value, then store the results into new list L3.Your command should look like the one at right. I have already carried out the command; the first few residuals are displayed. More can be seen by pressing [▸].

Alternatively, the TI-83 automatically finds residuals. They can be accessed [2nd][STAT] (LIST). Notice there is a list called RESID. These are the residuals from the *last* regression performed.

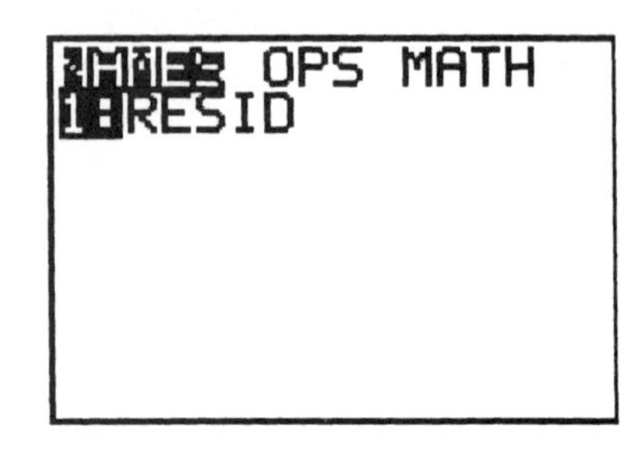

There are two main types of residuals plots which should be done to examine the adequacy of the model for any regression. The first plots the residuals against *X* (the predictor variable); the second is a normal probability plot of the residuals. In the first plot, we hope to see random scatter in an even band around the *X*-axis ($Y = 0$ line). Any departures from this are cause for reexamination of the model. In particular, curves may appear which are "masked" by the original scaling of the data; subtraction of any linear trend will magnify any curve. Another common shape which indicates problems is a "fan" in which plot either narrows from left to right or, alternatively, thickens. Either of the fan shapes means there is a problem with an underlying assumption: namely that the variation around the line is constant for all *x*-values. If this is the case, a transformation of either *Y* or *X* is usually necessary. Unusual observations (outliers) may also be seen in these plots as very large positive or negative residuals.

A residuals plot against *X* (the predictor variable)

This is a scatter plot. From the plot definitions screen (press [2nd][Y=][ENTER] to define Plot1) define the plot using the original *x*-list of the regression (in out example L1) and the residuals list (from our examples L3). Press [ZOOM][9] to display the graph.

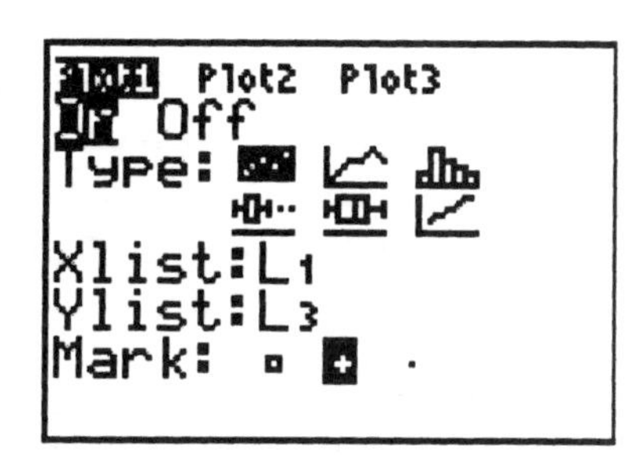

Alternatively, if using the automatic residuals, define the plot as above, but for the Ylist, press [2nd][STAT] and [ENTER]to select RESID, followed by [ZOOM][9]. The definition screen will look like the one at right.

In either case, the residual plot should look like the one at right. Looking at the plot, we see no overt curves, indicating a line appears to be an adequate model. This was a small data set; with these seeing non-constant variation can be difficult. There does not seem to be much of a problem except at the far right end of the graph, but there were few points there, so it's hard to tell. There do not seem to be any extremely large positive or negative residuals (outliers).

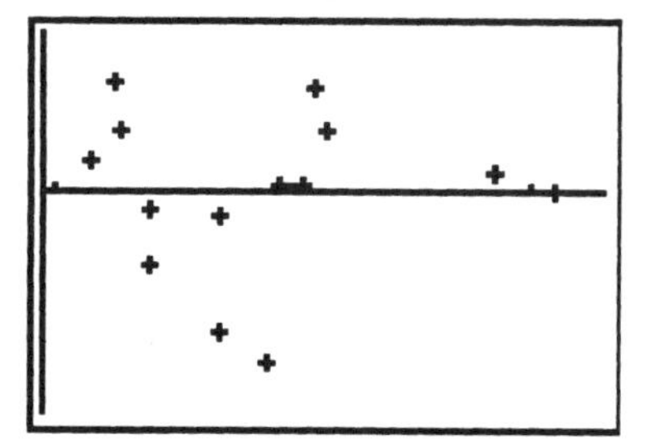

Normal Probability Plots of Residuals

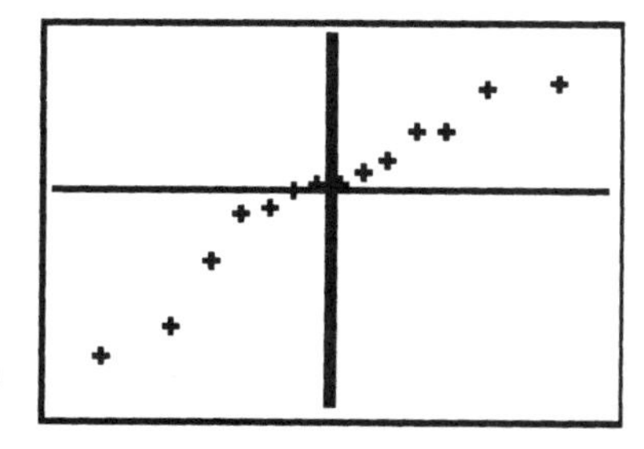

The second plot which should be done is a normal probability plot, since there is an underlying assumption the residuals have a normal distribution. This assumption will be used later in inference for regression. Normal probability plots were discussed in Chapter 4. Remember that the data list is the list of residuals. We're looking for a (approximately) straight line. Here, the pattern is (very) roughly linear, indicating no serious problems with this assumption.

Residuals Plots against Time

If the data were gathered through time (the data in our example were not) a time plot of the residuals should be done as discussed in Chapter 2. Ideally, this should look like random scatter. Any obvious patterns (lines, curves, fans, etc) indicate time is an important factor and the model which was fit is not adequate to fully describe the relationship. This generally means a multiple regression is needed to explain the response variable.

Identifying Influential Observations

Remember, the two points on the far right side of the original plot looked unusual. Points far away from the center of the range of the predictor variable can be influential; that is, they may have a significant impact on the slope, especially if they do not follow the pattern of the rest of the data. Even if they do not impact the slope, they will cause r and r^2 to be larger than the rest of the data would warrant. To decide if points are influential, delete the suspects, and reanalyze the data.

Here are the first few data values after having deleted the data for the Ford Excursion (310, 10) and GMC Yukon (285, 13).

L1	L2	L3 3
170	22	
190	15	
127	29	
215	21	
170	18	
194	21	
205	20	

L3(1)=

Redrawing the scatterplot (L1 as Xlist and L2 as Ylist) with [ZOOM][9] gives the plot at right. This looks much less linear than the original data scatterplot. One could conceivably think this is no longer "straight enough" for a linear regression. But let's try it anyway.

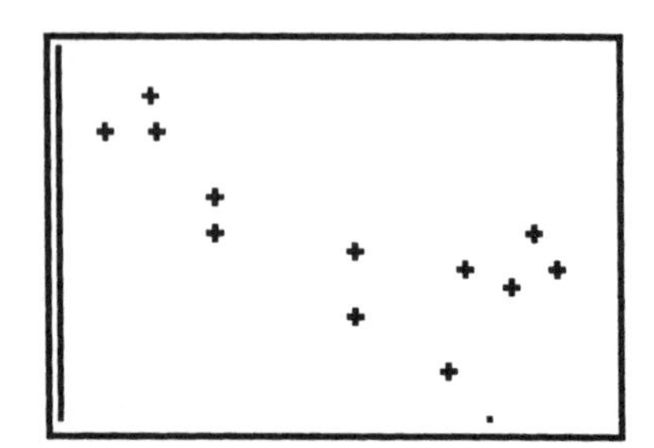

The new linear regression output is at right. The new regression equation is $Mileage = 39.39 - 0.10 * horsepower$. The slope changed only by about 10%, which is not much. Notice that r and r^2 are much less than before (as expected).

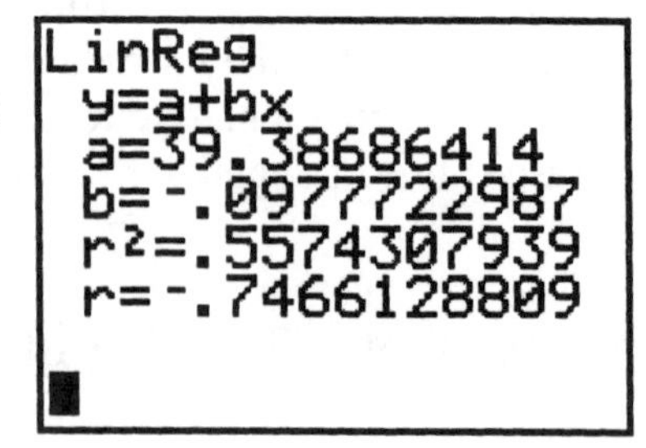

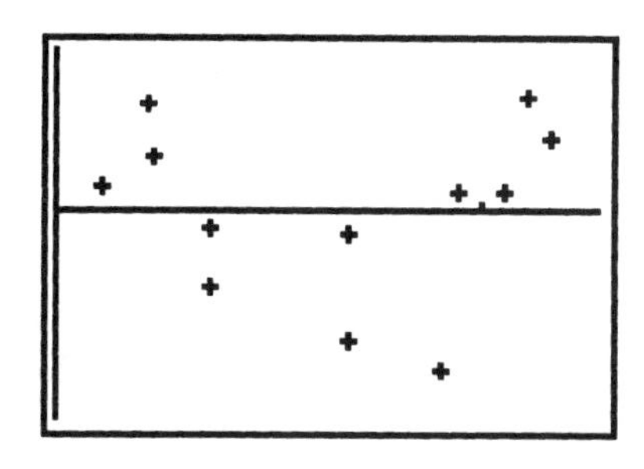

Here's a residuals plot against horsepower (*X*). We can definitely see the curve in these residuals. They're positive at both ends and negative in the middle.

We are left with the following indications: The Ford Excursion and GMC Yukon were influential in this regression; not so much for their impact on the regression equation, but because they made the data much more linear. With these two vehicles removed from the data set, the indication is that a line is *not* the proper model to describe the relationship between horsepower and gas mileage. What is correct? That is beyond the scope of this book; perhaps a model with an x^2 or x^3 term will be better. The TI-83 can calculate these regressions; if you wish they can be checked for adequacy by residuals plots just as we have done here. (It turns out that adding horsepower^2 makes a pretty good relationship.)

Transforming Data

There are two reasons to transform data in a regression setting: to straighten a curved relationship and to transform variability so it is constant around the line. In a single variable case, transformations can be used to make skewed distributions look more symmetric; in the case of a single variable observed for several groups, a transformation can make the different groups look more equally spread.

The table below shows stopping distances in feet for a car tested three times at each of five speeds. We hope to create a model that predicts stopping distance from the speed of the car.

Speed (mph)	Stopping Distance (ft)
20	64, 62, 59
30	114, 118, 105
40	153, 171, 165
50	231, 203, 238
60	317, 321, 276

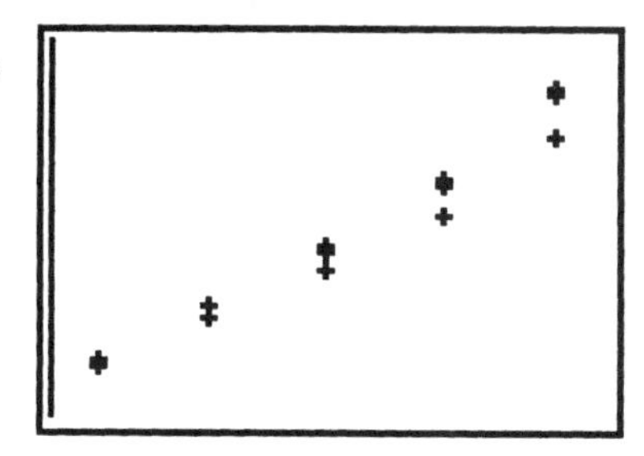

A plot of the data is at right. It looks fairly linear, but it is clear that the stopping distances become more variable with faster speed.

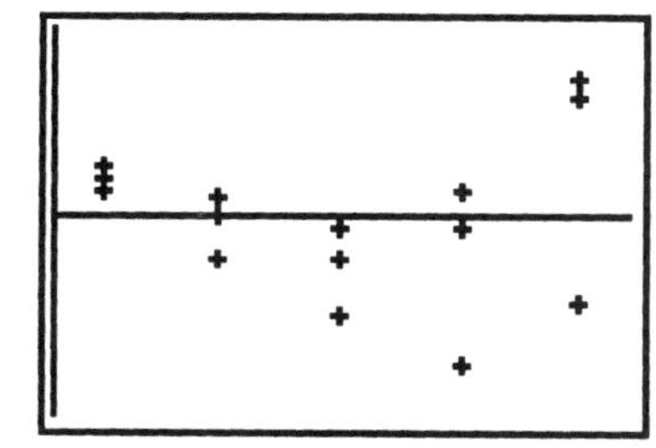

Regression gives the fitted model as $Stoppingfeet = -65.933 + 5.98 * Speed$. The residuals plot against speed (at right) clearly indicates the variability gets larger for faster speeds; it also indicates the true relationship is not linear but curved. Clearly, a transform is indicated – one which will decrease variation as well as straighten the plot.

Since the residuals indicate a curve (possibly quadratic), using the square root of stopping distance makes sense. With stopping distance in L2, we need to find the square root of each distance. With one command we can do this, storing the result in a new list, say L3. Press [2nd][x^2] ($\sqrt{\ }$) [2nd][2](L2)[)] [STO▸][2nd][3] followed by [ENTER]. The command and result are at right. We see the first few values. To see the entire list, go to the Stat editor.

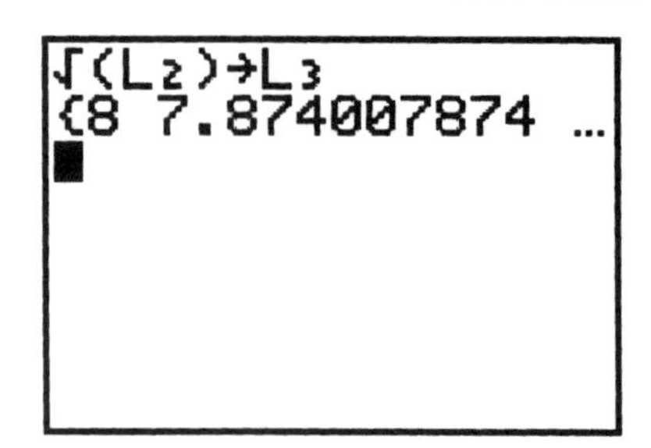

The new scatterplot is at right. The new regression equation is $sqrt(StopDist) = 3.303 + 0.235 * Speed$. We have $r = 0.9922$, an extremely strong linear relationship. What about a residuals plot?

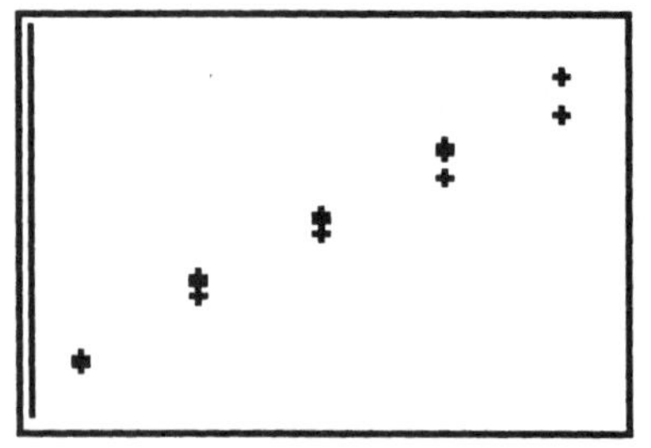

Here is the residuals plot. It's not perfect; the variation still increases with larger values of speed, but is much better than before. Sometimes there is no "perfect" transform.

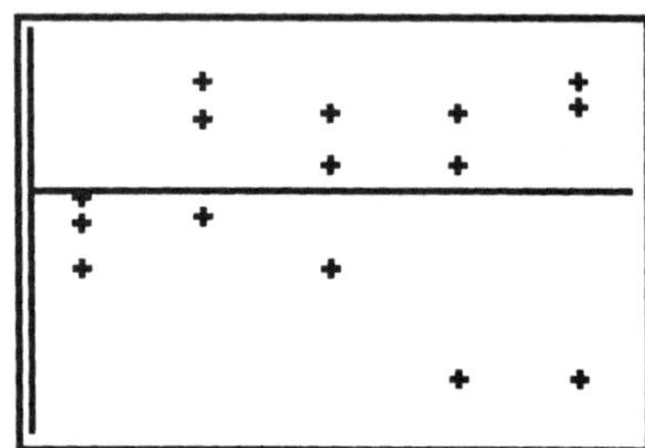

What can go wrong?

What's Dim Mismatch?

We've seen this one before. Press [ENTER] to quit. This error means the two lists referenced (either in a plot or a regression command) are not the same length. Go to the Stat editor and fix the problem.

ERR:DIM MISMATCH
1:Quit

Err: Invalid?

This error is caused by referencing the function for the line when it has not been stored. Recalculate the regression being sure to store the equation into a *Y* function.

What's that weird line?

This error can come either in a data plot (an old line still resides in the [Y=] screen) or the stored regression line is showing in the residuals plot, as shown here. The regression line is not part of the residuals plot and shows only because the calculator tries to graph everything it knows about. Press [Y=] followed by [CLEAR] to erase the equations, then redraw the graph by pressing [GRAPH].

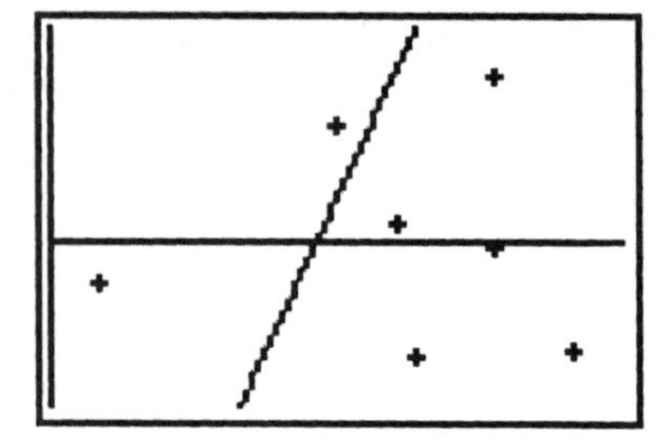

Nonreal Ans?

This error comes from trying to take the square root (or log) of a negative number. Sorry, can't be done in the real number system. These transforms do not work for negative values. Try something else.

ERR:NONREAL ANS
1:Quit
2:Goto

This doesn't look like a residuals plot!

It doesn't. Residuals plots *must* be centered around $Y = 0$. This error is usually caused by confusing which list contains the Y's and which the X's in finding residuals by "hand." Go back and check which list is which, then recomputed the residuals, or use the list automatically stored by the calculator (under the LIST menu.)

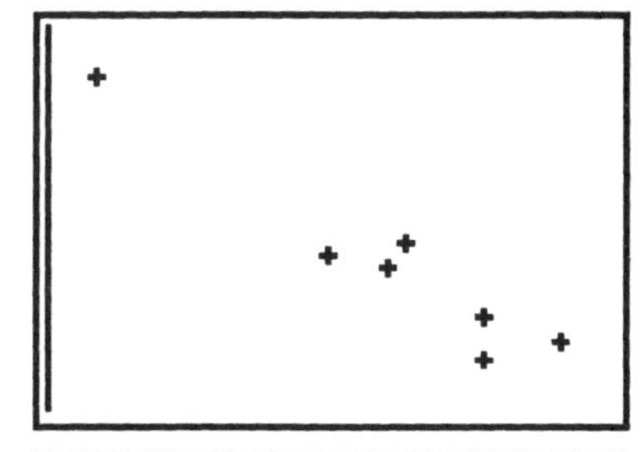

Invalid?

This is another error that occurs when explicitly finding residuals. The problem here was that the regression equation was not stored into a function when it was calculated. Redo the regression, being sure to include Y_1 as the last parameter in the command.

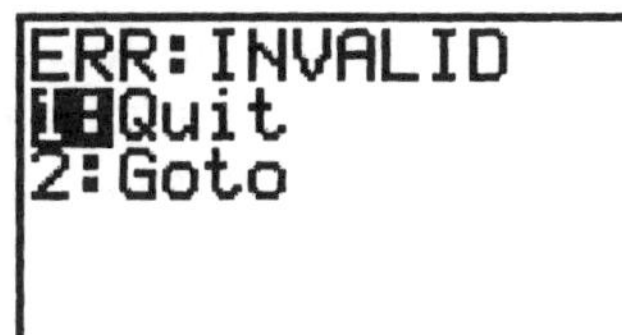

Chapter 6 – Random Numbers

Randomness is something most people seem to have an intuitive sense about. But truly random values are surprisingly hard to get. In fact, calculators (and computers) can't generate true random numbers since any values they obtain are based on an algorithm (that is, a program). But they do give good *pseudorandom* numbers. These have many applications from simulation to selecting samples and assigning treatments in an experiment.

Simulations

Simulations are used to mimic a real situation such as this. Suppose a cereal manufacturer puts pictures of famous athletes in boxes of their cereal as a marketing ploy. They announce that 20% of the boxes contain a picture of Tiger Woods, 30% a picture of Lance Armstrong, and the rest have a picture of Serena Williams. How many boxes of the cereal do you expect to have to buy in order to get a complete set?

You could go out and buy lots of cereal, but that might be expensive. We'll model the situation using random numbers, assuming the pictures really are randomly placed in the cereal boxes, and distributed randomly to stores across the country.

We'll use random digits to represent getting the pictures: Since 20% have Tiger's picture, we'll let the digits 0 and 1 represent getting his picture. Similarly, we'll use digits 2, 3, and 4 (30% of the 10 digits) to represent getting Lance's picture. The rest (5 through 9) will mean we got a picture of Serena.

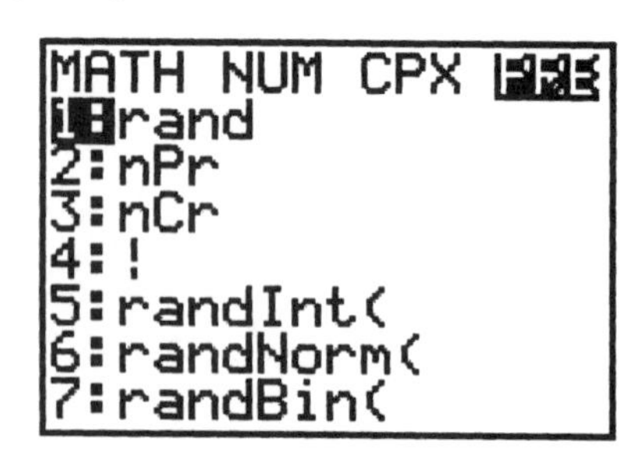

We need to get the random digits. From the home screen, press [MATH], then arrow to PRB. The menu at right is displayed. We want choice 5:randInt(. Either arrow to it and press [ENTER] or simply press [5]. The command shell will be transferred to the home screen. Now you need to tell the calculator the boundary values you want. Since we want digits between 0 and 9, we will enter [0][,][9] . Pressing [ENTER] will get the first random digit.

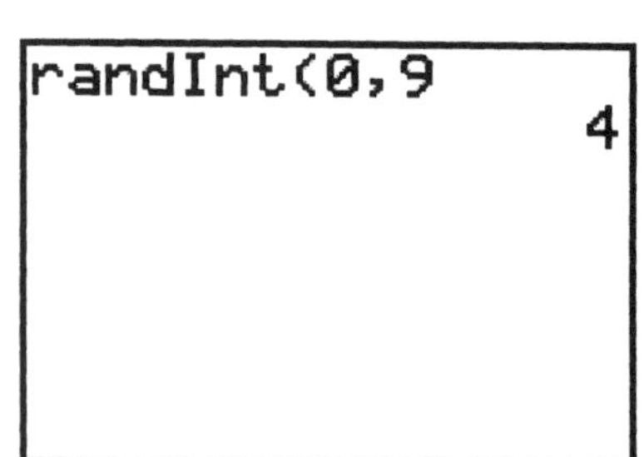

Here is our first random digit: a 4. That means the first box had a picture of Lance. We can continue pressing [ENTER] and get more random digits.

Controlling the sequence of Random Digits

You didn't get the same random number? Not surprising. Random number generation on computers and calculators works from something called a *seed.* In the case of TI calculators, every command you use changes the seed. If a value is explicitly stored as the seed *immediately before* a random number command, the sequence of random digits will be the same every time.

To store a seed, enter the value desired, then press [STO▶][MATH], then arrow to PRB and press [ENTER] to select rand then press [ENTER] again to actually store the seed. The sample at right stored 1187 as the seed.

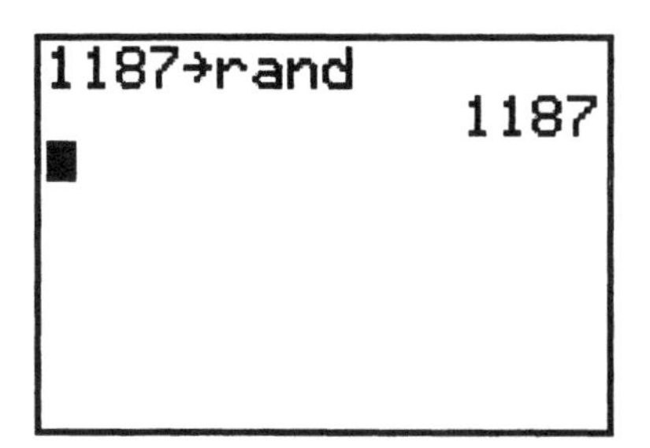

Following this with the same random integer command as before, we get values as shown at right. Yours should be the same. Look at the first five digits. These correspond to (in our example above) getting Serena, Serena, Serena, Serena, Tiger, and Serena. Even after the sixth "box" we haven't gotten all three pictures. In fact, it takes two more boxes (another Tiger then finally a Lance) for a total of eight boxes to get the complete set.

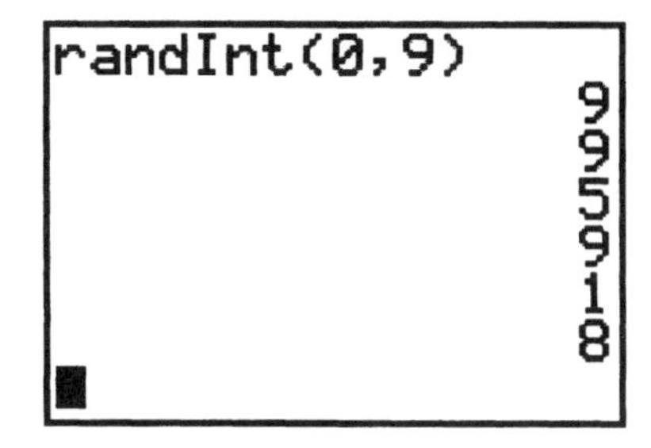

One simulation is not a very good representation. We'd like to know how many boxes it would take to get all three, *on average*. We need to repeat the simulation many times, and take the average value from the many simulations. We could just keep pressing [ENTER] until we've done enough, or we can get many random numbers at once and store them into a list.

Here, we've reset the seed to 18763 and changed the random integer command to add another parameter – how many numbers to generate. Since it could possibly take many boxes of cereal to get all three pictures, we've chosen to store 200 numbers into list L1. (Press [STO▶][2nd][1] after the ending parenthesis on the random integer command.) The first few values are displayed. To see the rest, use the Statistics Editor.

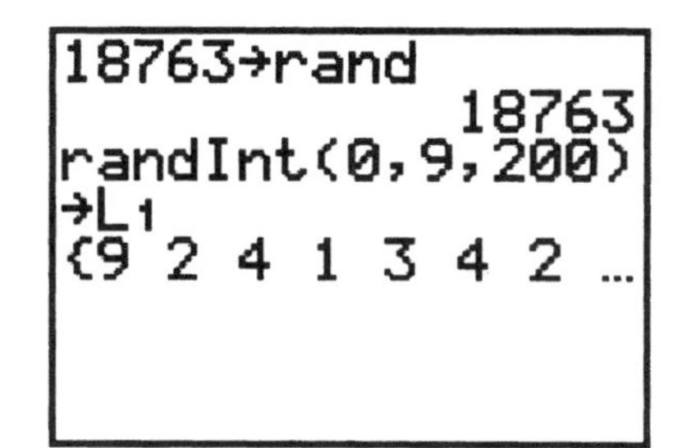

Looking at the list in the editor, the first four digits are 9, 2, 4, and 1. That corresponds to a Serena, Lance, Lance, and Tiger. That trial took four boxes to get the full set. One can continue down the list until several complete sets have been found; then compute the average for all trials as the estimate of the average number of boxes.

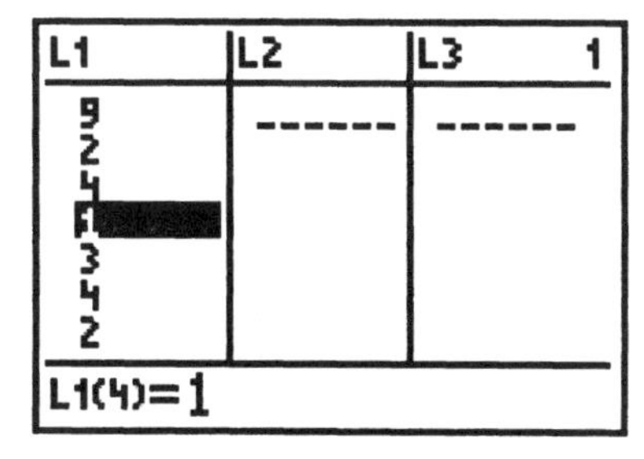

Random Normal Data

These calculators can also simulate observations from normal populations in a manner similar to the examples above. The command is choice 6: randNorm(from the Math, Probability menu. The parameters are the mean and standard deviation.

This example models the following: A tire manufacturer believes that the tread life of their snow tires can be described by a Normal model with mean 32,000 miles and standard deviation 2500 miles. You buy 4 of these tires, hoping to drive them at least 30,000 miles. Estimate the chances that all four last at least that long. We have output for one trial – a set of four tires. In this trial, 3 of the 4 lasted over 30000 miles. To further estimate the chance that all four last over 30,000 miles, obtain more repetitions of sets of four tires. Notice that the mileages are entered in the command without commas in the large numbers. If you tried to use a comma, the calculator would interpret as a parameter separator.

```
randNorm(32000,2
500
     31114.06831
      27790.6048
     30573.20062
     32507.09429
```

Sampling and Treatment Assignments

Random numbers are the best method for (randomly!) selecting items or individuals to be sampled or treatments to be assigned in an experiment. In the sampling frame (a list of members of the population) number the individuals from 1 to N, where N is the total number in the list. Use the Random integer command to select those to be sampled. In the case of assigning treatments, if there are, for example two treatments, use random integers to assign half the experimental units to treatment A; the rest will get treatment B.

Chapter 7 – Probability Models

We've already used the calculator to find probabilities based on normal models. There are many more models which are useful. This chapter explores two such models.

Many types of random variables are based on Bernoulli trials experiments. These involve independent trials, only two outcomes possible, and a constant probability of success called *p*. Two of the more common of these variables have either a Geometric or Binomial model.

Geometric Models

The Geometric probability model is used to find the chance the first success occurs on the nth trial. If the first success is on the nth trial, it was preceded by *n-1* failures. Because trials are independent we multiply the probability of a failure, 1-p, times itself n-1 times then multiply by the probability of a success, so we have $P(X = n) = (1-p)^{n-1}p$ which is sometimes written as $P(X = n) = (q)^{n-1}p$. This is generally easy enough to find explicitly, but the calculator has a built-in function to find this quantity as well as the probability the first success comes somewhere on or before the nth trial.

Suppose we are interested in finding blood donors with O-negative blood; these are called "universal donors." Only about 6% of people have O-negative blood. In testing a group of people, what is the probability the first O-negative person is found on the 4th person tested? We want $P(X = 4)$. Press [2nd][VARS] (DISTR). We want menu choice D:geometpdf(. To find the menu option, you can press the down arrow until you find it, press [ALPHA][x⁻¹](D) , or press the up arrow twice. The last is probably the easiest. Press [ENTER] to select the option. The command shell is transferred to the home screen.

```
DISTR DRAW
9↑Fcdf(
0:binompdf(
A:binomcdf(
B:poissonpdf(
C:poissoncdf(
D:geometpdf(
E:geometcdf(
```

Enter the two parameters for the command: p and n. Here, p=.06 and $n = 4$. Enter them separated by a comma, then press [ENTER] to find the result. We see there is about a 5% chance to find the first O-negative person on the 4th person tested (assuming of course that the individuals being tested are independent of each other.)

```
geometpdf(.06,4)
              .04983504
```

There are some other related questions that can be asked. What is the probability the first O-negative person will be found somewhere in the first 5 persons tested? We want to know $P(X \le 5)$. We could find all the individual probabilities for 1, 2, 3, 4, and 5 and add them together but there is an easier way. We really want to "accumulate" all those probabilities into one, or find the *cumulative* probability. This uses menu choice E:geometcdf(. The parameters are the same as before, namely p and n. Finding the first O-negative person within the first 5 people tested should happen about 26.6% of the time.

```
geometpdf(.06,4)
              .04983504
geometcdf(.06,5)
            .2660959776
```

```
                .04983504
geometcdf(.06,5)
               .2660959776
1-geometcdf(.06,
9)
               .5729948022
```

What's the chance we'll have to test at least 10 before we find an O-negative person? We want $P(X \geq 10)$. Notice from above that the cdf command gave the probability the random variable is less than or equal to some value. We can make use of that fact and the idea of complements. The event $X \geq 10$ is the opposite of $X < 10$, which here is really $X \leq 9$, so we will find $P(X \geq 10)$ as $1 - P(X \leq 9)$ as in the screen at right. Notice that starting the command with [1][-] then going to the distributions menu allows using only a single command. Alternatively, I could have found the geometric probability and then subtracted the result from 1. There is a 57.3% chance we'll have to test at least 10 people before finding an O-negative person.

An important note: When finding probabilities of this type for a "greater than/more than" or "at least" setting, always find the cumulative probability (cdf) for the high end of what is not wanted and then subtract that result from 1.

Binomial Models

Binomial models are interested in the chance of k successes occurring when there are a fixed number (n) of Bernoulli trials.

Consider a family planning on five children. What is the chance they will have three girls? Girls and boys do not actually each happen half the time. Boys are really born about 51.7% of the time. This means girls are born 48.3% of the time. We know pregnancies are independent of each other (unless multiple births are involved). In order to have three girls, the family could have girl, girl, boy, girl, boy or boy, boy, girl, girl, girl, or any one of several different arrangements, each of which will have the same probability. How many arrangements are possible so that there are 3 girls in 5 children? The Binomial coefficient provides the answer to this question. The coefficient itself is variously written as $\binom{n}{k}$ or nCk and is read as "n choose k." To finally answer our question, type the number 5 on the home screen, then press [MATH] arrow to PRB and select choice 3:nCr then type in 3 and press [ENTER]. We find there are 10 possible arrangements of three girls in a family of five children. We can then find the probability of three girls in a family of five children as $10(.517)^2(.483)^3 = 0.301$. This means about 30% of families with five children should have three girls.

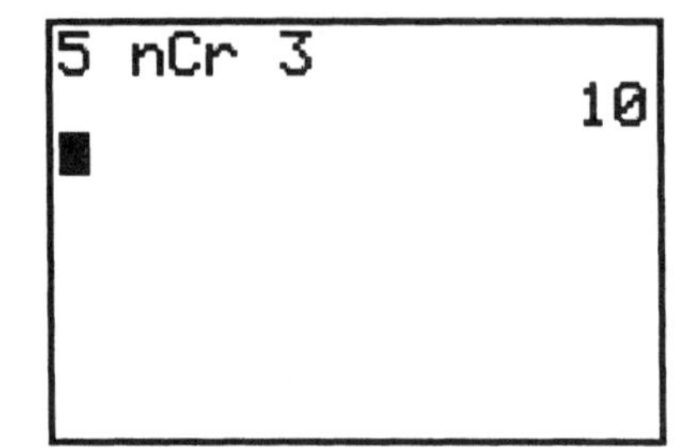

There is an easier way to find these binomial probabilities. Looking back at the portion of the DISTR menu shown on the previous page there are two menu options that will help us here: 0:binompdf(and A:binomcdf(. Just as in the case of the geometric model discussed above, the pdf menu choice gives $P(X = k)$ and the cdf gives $P(X \leq k)$. Three examples follow.

Press [2nd][VARS] (DISTR). Either arrow down to choice 0 and press [ENTER], press 0, or press the up arrow until you find choice 0 and press [ENTER]. The command shell is transferred to the home screen. There are three parameters to enter separated by commas: n (the number of trials), p (the probability of success), and k (the value of interest).

Here we have entered the command and pressed [ENTER] to find the result. This is just the quantity we calculated explicitly above, namely there is about a 30% chance that a family with five children will have three girls.

```
binomPdf(5,.483,
3)
         .3011774684
```

Returning to the prior example about blood donors, what is the probability that if 20 donors come to the blood drive, there will be 3 O-negative donors? From the screen at right, we see this is 8.6%.

```
binomPdf(5,.483,
3)
         .3011774684
binomPdf(20,.06,
3)
         .0860066662
```

As with geometric models, we can find the use the cdf command to find cumulative probabilities. Remember, the calculator adds individual terms for $X = 0, 1, \ldots$ and so on up to the upper limit specified. For instance, what is the chance of at most 3 O-negatives in a blood drive with 20 donors? We see this is very likely to happen. We'll expect 3 or fewer O-negative donors in 20 people about 97% of the time.

```
         .3011774684
binomPdf(20,.06,
3)
         .0860066662
binomcdf(20,.06,
3)
         .9710342619
```

What's the chance there would be more than 2 O-negative donors in a group of 20? Again, since we are looking for P($X > 2$) and looking toward the high end, so this will be found with by subtracting the cdf results for the *high end of what is not wanted* from 1. We want everything above 2, so 2 is the high end of the unwanted portion of the distribution. Notice the command as entered does the entire calculation at once. If desired, one could first find the probability of 2 or less, and then subtract that result from 1. In a group of 20 donors, we'll get more than 2 O-negative people about 11.5% of the time.

```
         .0860066662
binomcdf(20,.06,
3)
         .9710342619
1-binomcdf(20,.0
6,2)
         .1149724038
```

A final word about binomials. Some older versions of these calculators (and computer applications as well) cannot deal with large values of n. This is because the binomial coefficient becomes too large very quickly. Newer models use an approximation. However, when n and p are sufficiently large (generally, both of $np \geq 10$ and $n(1-p) \geq 10$ must be true to move the distribution away from the ends so it can become symmetric) binomials can be approximated with a normal model. One uses the normalcdf command described in Chapter 4 specifying the mean as the mean of the binomial ($\mu = np$) and the standard deviation as that of the binomial ($\sigma = \sqrt{np(1-p)}$).

Suppose the Red Cross anticipates the need for at lease 1850 units of O-negative blood this year. They anticipate having about 32,000 donors. What is the chance they will not get enough O-negative blood? We desire P($X < 1850$). We have calculated the mean to be 1920 and the standard deviation to be 42.483. Practically speaking the low end of interest is 0, but remember the normal distribution extends to $-\infty$. It appears there is about a 5% chance there will not be enough O-negative in the scenario discussed.

```
√(32000*.06*.94)
          42.48293775
normalcdf(-9999,
1850,1920,42.483
)
          .049705269
```

What can go wrong?

Err:Domain?

This error is normally caused in these types of problems by specifying a probability as a number greater than 1 (in percent possibly instead of a decimal). Reenter the command giving p in decimal form.

```
ERR:DOMAIN
1:Quit
2:Goto
```

How can the probability be more than 1?

It can't. As we've said before, if it looks more than 1 on the first glance, check the right hand side. This value is 3.08×10^{-6} or 0.000003, not 3.08.

```
,6)
          .0033431611
binompdf(25,.517
,6.5)
binompdf(25,.017
,6)
     3.086234546E-6
```

Chapter 8 – Inference for Proportions

We know the sample proportion, $\hat{p}$, is normally distributed if both $np \geq 10$ and $n(1-p) \geq 10$. With this fact we can obtain probabilities based on the normal model of obtaining certain sample proportions. Inference asks a different question. Based on a sample, what can we say about the true population proportion? Confidence intervals give values of believable results along with a probability statement giving our level of certainty that the interval contains the true value. Hypothesis tests are used to decide if a claimed value is or is not reasonable based on the sample.

Confidence Intervals for a single proportion

Sea fans in the Caribbean Sea have been under attack by a disease called *aspergillosis*. Sea fans that can take up to 40 years to grow can be killed quickly by this disease. In June 2000, members of a team from Dr. Drew Harvell's lab sampled sea fans at Las Redes Reef in Akumai, Mexico at a depth of 40 feet. They found that 54 of the 104 fans sampled were infected with the disease. What might this say about the prevalence of the disease in general? The observed proportion, $\hat{p} = 54/104 = 51.9\%$ is a point estimate of the true proportion, p. Other samples will surely give different results.

We can use the calculator to obtain a confidence interval for the true proportion of infected sea fans. Press STAT then arrow to TESTS. The first portion of the menu shows several hypothesis tests. We'll talk about them later. Arrowing down, we come to several possible intervals. The one we want here is choice A:1-PropZInt. Either arrow to it and press ENTER or press ALPHA MATH (A).

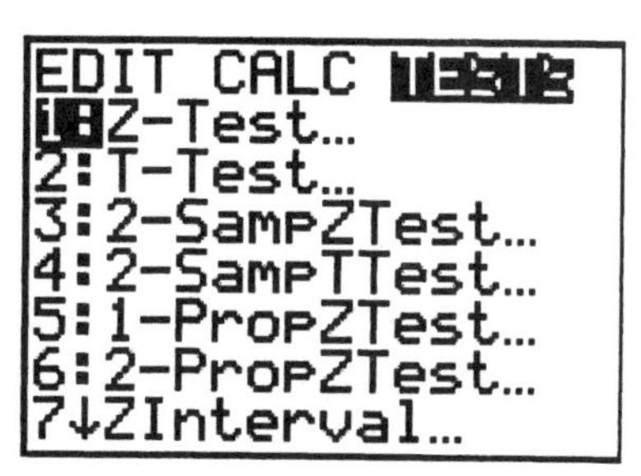

Here is the input screen. Simply enter the number of observed "successes," which here is the number of infected sea fans, 54, press the down arrow, then enter the number of trials, 104 fans were observed, press the down arrow again to enter the desired level of confidence, finally press the down arrow again and press ENTER to calculate the results.

1-PropZInt
x:54
n:104
C-Level:.95
Calculate

How much confidence? That is up to the individual researcher. The trade-off is that more confidence requires a wider interval (more possible values for the parameter). 95% is a typical value, but the level is generally specified in each problem.

Here are the results. The interval is 0.423 to 0.615. Remember the calculator usually gives more decimal places than are really reasonable. Usually reporting proportions to tenths of a percent is more than enough. Your instructor may give other rules for where to round final answers. The output also gives the sample proportion and the sample size.

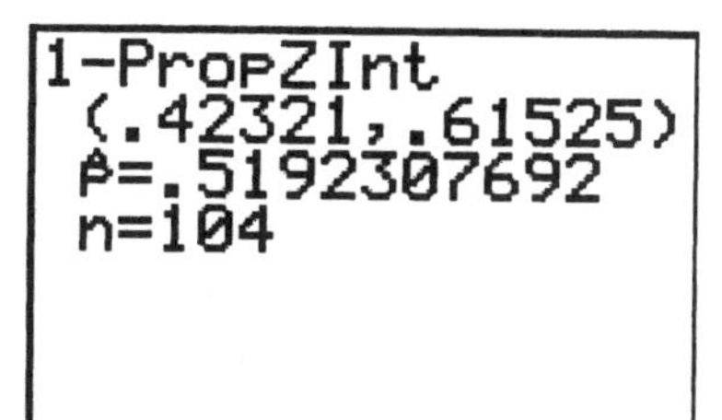

What can we say about the prevalence of disease in sea fans? Based on this sample, we are 95% confident that between 42.3% and 61.5% of Las Redes sea fans are infected by the disease.

In August 2000, the Gallup poll asked 507 randomly sampled adults the question "Do you think the possession of small amounts of marijuana should be treated as a criminal offense?" Of these, 47% answered "No." What can we conclude from the survey?

Results from polls are usually given in (rounded) whole percents. In order to create a confidence interval using the calculator we need the number of respondents who answered "No." Multiplying 0.47*507 gives 238.29. Since there can't be a fraction of a response, round this number to 238 "No" responses. The input screen is at right.

```
1-PropZInt
 x:238
 n:507
 C-Level:.95
 Calculate
```

Based on this poll, we can be 95% confident the proportion of Americans who think possession of small amounts of marijuana should not be treated as a criminal offense is between 42.6% and 51.3%.

```
1-PropZInt
 (.42599,.51287)
 p̂=.4694280079
 n=507
```

Hypothesis tests for a single proportion

Confidence intervals give ranges of believable values for the parameter (in this case the proportion of successes.) Hypothesis tests assess the believability of a claim about the parameter. Certainly, if a claimed value is contained in a confidence interval it is plausible. If not, it is unreasonable. Formal tests of hypotheses assess the question somewhat differently. The results given include a test statistic (here a z-value based on the standard model) and a p-value. The p-value is the probability of a sample result as or more extreme, given the claimed value of the parameter. Large p-values argue in support of the claim, small ones argue against it; in essence, if the claimed value were true the likelihood of observing what was seen in the sample is very small.

In some cultures, male children are valued more highly than females. In some countries with the advent of prenatal tests such as ultrasound, there is a fear that some parents will not carry pregnancies of girls to term. A study in Punjab India[1] reports that in 1993 in one hospital 56.9% of the 550 live births were males. The authors report a baseline for this region of 51.7% male live births. Is the sample proportion of 56.9% evidence of a change in the percentage of male births?

Press [STAT] and arrow to TESTS. Select choice 5:1-PropZTest by either pressing [5] or arrowing to the selection and pressing [ENTER]. We are asked for p_0, the posited value which is 51.7% in this case. Enter the proportion as the decimal 0.517. Then we need the number of "successes" (Multiplying 0.569*550 gives 312.95 which rounds to 313). The multiplication can be entered directly by the x: prompt, then press [ENTER], then uparrow to round the result if needed. The number of trials, n, is 550. Then we need a direction for the alternate hypotheses (what we hope to show). This is usually obtained from the form of the question. In this case, we want to know if the proportion has changed which might argue for selecting the

[1] "Fetal Sex determination in infants in Punjab, India: correlations and implications", E.E. Booth, M. Verna, R. S. Beri, *BMJ*, 1994; 309:1259-1261 (12 November).

$\neq p_0$ alternative, but we suspect that this proportion should increase from the baseline if male births are being selected. Move the cursor with the right arrow to highlight the alternative $> p_0$ and press [ENTER] to move the highlight. Finally, there are two choices for output. Selecting Calculate merely gives the results. Selecting Draw draws the normal curve and shades in the area corresponding to the p-value of the test. The input screen for the test is at right.

```
1-PropZTest
 p0:.517
 x:313
 n:550
 prop≠p0 <p0 >p0
 Calculate Draw
```

At right is the output from selecting Calculate. The first line of the output gives what the calculator understood the alternate hypothesis to be. Always check that this is what you intended, as it can make a difference in the p-value. The value of the test statistic is $z = 2.44$. This means that if there were no change, the observed 56.9% is 2.44 standard deviations above the mean. The p-value for the test is 0.0072. This means that if the proportion of male births is still 51.7%, we would observe a value of 56.9% or greater only about 7 times in 1000. Since this is very rare, we will reject the null hypotheses and conclude that we believe that, based on these data, the true proportion of male births in Punjab is now greater than the baseline 51.7%.

```
1-PropZTest
 prop>.517
 z=2.444693651
 p=.0072487641
 p̂=.5690909091
 n=550
```

Here is the output when DRAW is selected. There is not as much information given, but the test statistic and p-value are reported. Since this p-value is so small, not much is shown as shaded.

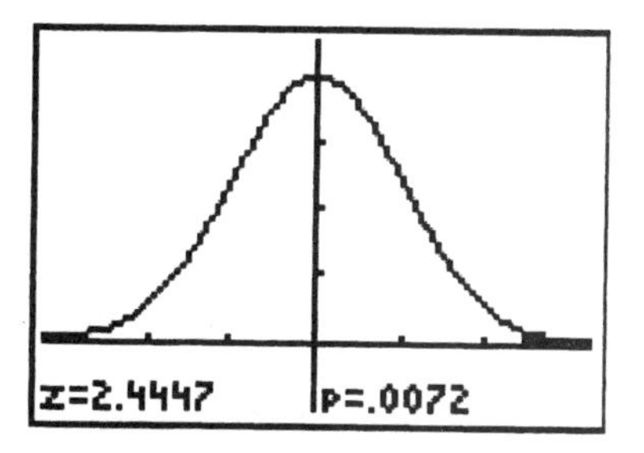

When a test rejects the null hypothesis as in the example above (remember, we decided the proportion of male births is now more than the baseline) it is good practice to report a confidence interval for where the parameter is, based on the sample. Find a 95% confidence interval for the true proportion of male births in Punjab, as detailed above. Notice that the data entry in the input screen is preserved from the hypothesis test. Based on these data, we are 95% confident the true proportion of male births in Punjab is now between 52.8% and 61.0%.

```
1-PropZInt
 (.52771,.61048)
 p̂=.5690909091
 n=550
```

Is there a home field advantage in baseball? In 2002, the home team won 1314 of 2425 games. That's 54.2%. It's more than the 50% we would expect if there were no home field advantage, but is it enough bigger to say it's statistically significant? The input screen is shown at right. Notice that if there is a home team advantage, we'd expect the home team to win more than 50% of games, so the alternate is $> p_0$.

```
1-PropZTest
 p0:.5
 x:1314
 n:2425
 prop≠p0 <p0 >p0
 Calculate Draw
```

Here are the results of the test. The z-statistic of 4.12 which means if there is no home field advantage, the observed 54.2% is 4 standard deviations above the mean, which is extremely rare. The p-value is (be careful here!) 1.88×10^{-5} or 0.0000188. This extremely small p-value argues that these data show there is indeed a home field advantage.

```
1-PropZTest
 prop>.5
 z=4.12230543
 p=1.8765481E-5
 p̂=.5418556701
 n=2425
```

How large is the advantage? Based on these data, we are 95% confident the home team will win between 52.2% and 56.2% of baseball games.

```
1-PropZInt
 (.52203,.56169)
 p̂=.5418556701
 n=2425
```

Confidence Intervals for the difference in two proportions

Who are typically more intelligent, men or women? To find out what people think, the Gallup Poll sampled 520 women and 506 men. They showed them a list of attributes and asked them to indicate whether each attribute was "generally more true about men or women."[2] When asked about intelligence, 28% of men thought men were normally more intelligent, but only 14% of women agreed. The difference is 14%, which looks large, but is it large enough to be meaningful or is it due to sampling variability? We will first compute a confidence interval for the difference in the true proportions, $p_M - p_W$. If this interval contains 0, there is no statistical evidence of a difference.

From the STAT, TESTS menu select choice B:2-PropZInt... by either arrowing to it and pressing [ENTER] or by pressing [ALPHA][MATRX] (B). The calculator uses groups 1 and 2 not men and women and calculates results based on $group1 - group2$. Decide (and keep note of) which group you call which. Since it is usually easier to deal with positive numbers, we will call the males group 1. As before, we calculate numbers of "successes" for men as 0.28*506 = 141.68 (round this to 142) and for women as 0.14*520 = 72.8 (which rounds to 73). The input screen should look like the one at right.

```
2-PropZInt
 x1:142
 n1:506
 x2:73
 n2:520
 C-Level:.95
 Calculate
```

The calculated interval is 0.091 to 0.189. This means we are 95% confident, based on this poll, the proportion of men who think men are more intelligent is between 9.1% and 18.9% more than the proportion of women who think men are more intelligent. Since the interval does not contain 0, there is a definite difference in the two genders.

```
2-PropZInt
 (.09101,.18948)
 p̂1=.2806324111
 p̂2=.1403846154
 n1=506
 n2=520
```

[2] http://www.gallup.com/poll/releases/pr010221.asp

Hypothesis tests for a difference in proportions

The National Sleep Foundation asked a random sample of 1010 U.S. adults questions about their sleep habits. The sample was selected in the fall of 2001 from random telephone numbers.[3] Of interest to us is the difference in the proportion of snorers by age group. The poll found that 26% of the 184 people age 30 or less reported snoring at least a few nights a week; 39% of the 811 people in the older group reported snoring. Is the observed difference of 13% real or merely due to sampling variation?

The null hypothesis is there is no difference, or $p_1 - p_2 = 0$. (The calculator and most computer statistics packages can only test assumed differences of 0; if there were an assumed difference, say the belief is that older people had 10% more snorers than young people, one would need to compute the test statistic "by hand.")

Decide which group will be group1. We will use the older people as group 1. (There will be no difference in the results, but again, positive numbers are generally easier for most people to deal with.) From the STAT, TESTS menu select choice 6:2-PropZTest. The number of snorers in the older group is 0.39*811 = 316.29 (rounded to 316); for the younger group the number of snorers is 0.26*184 = 47.84 which rounds to 48. The chosen alternate is $\neq p2$ since we just want to know if there is a difference.

```
2-PropZTest
 x1:316
 n1:811
 x2:48
 n2:184
 p1:≠p2 <p2 >p2
 Calculate Draw
```

Here are the results. Be careful here, as there are lots of p's floating around. We first see the chosen alternative, $p1 \neq p2$, The value of the test statistic is z =3.27, which means that if there were no difference in the proportion of snorers the observed difference (about 13%) is more than 3 standard deviations above the mean. The p-value for the test is given next: p=0.0011. The next values given are the observed proportions in each group, $\hat{p}_1$ and $\hat{p}_2$ then an overall $\hat{p}$ which represents the observed proportion, *if there were no difference in the groups*. One can arrow down to see the sample sizes for the two groups as well. Since the p-value for the test is so small, we believe there is a difference in the rate of snorers based on this poll. We can further say the proportion of snorers is greater in older people than in those under 30.

```
2-PropZTest
 p1≠p2
 z=3.274084022
 p=.0010601709
 p̂1=.3896424168
 p̂2=.2608695652
↓p̂=.3658291457
```

How big is the difference? We are 95% confident, based on this data the proportion of snorers in older adults is between 5.7% and 20.1% larger than for those under 30.

```
2-PropZInt
 (.057,.20055)
 p̂1=.3896424168
 p̂2=.2608695652
 n1=811
 n2=184
```

[3] 2002 *Sleep in America Poll*, National Sleep Foundation. Washington D.C.

What can go wrong?

Err: Domain?

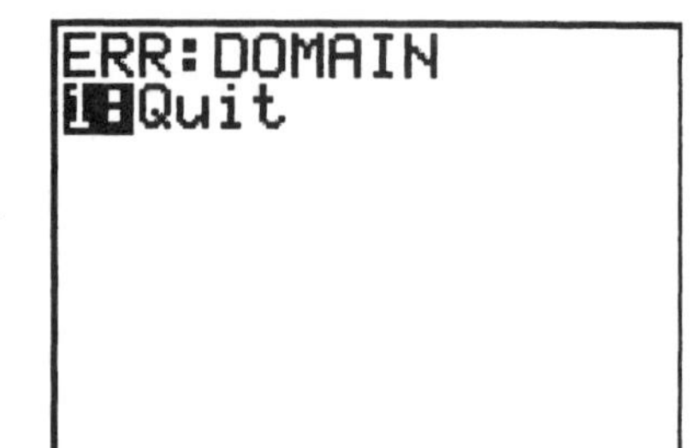

This error stems from one of two types of problems. Either a proportion was entered in a 1-PropZTest which was not in decimal form or the numbers of trials and/or successes was not an integer. Go back to the input screen and correct the problem.

Err:Invalid Dim?

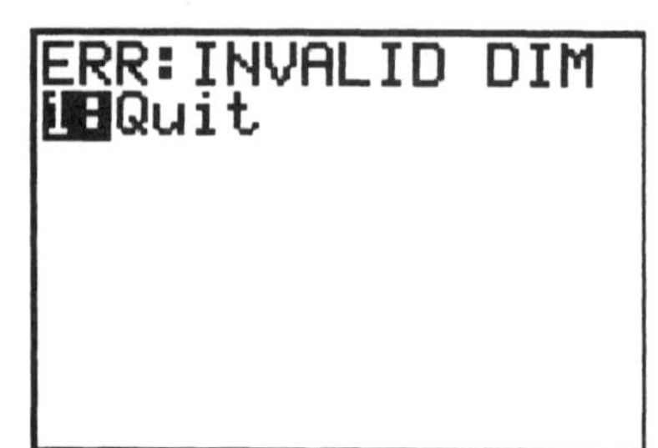

This can be caused by selecting the DRAW option if another Statistics plot is turned on. Either go to the STAT PLOT menu ([2nd][Y=]) and turn off the plot or redo the test selecting CALCULATE.

Bad Conclusions.

Small p-values for the test argue against the null hypothesis. If the p-value is small, one rejects the null hypothesis and believes the alternate is true. If the p-value is large, the null hypothesis is not rejected; this does *not* mean it is true – we simply haven't gotten enough evidence to show it's wrong. Be careful when writing conclusions to make them agree with the decision.

Chapter 9 – Inference for Means

Inference for means is a little different than that for proportions. Most introductory statistics texts base this on standard normal models which is truly appropriate only if the population standard deviation, σ, is known. In most cases this is not true; the only time one might really believe σ is known is in the case of quality control sampling where a production line has been tracked for a long time. If σ is not known confidence intervals and hypothesis tests should be based on t distributions. These become the standard normal distribution when the sample size is very large (infinite).

Small sample sizes give rise to their own problems. If the sample size is less than about 30, the Central Limit Theorem does not apply, and one cannot assume the sample mean has a normal distribution. In the case of small samples, you must check that the data come from a (at least approximately) normal population, usually by normal probability plots since histograms are not useful with small samples.

Confidence Intervals for a mean

Residents of a small northeastern town who live on a busy street are concerned over vehicles speeding through their area. The posted speed limit is 30 miles per hour. A concerned citizen spends 15 minutes recording the speeds registered by a radar speed detector that was installed by the police. He obtained the following data:

29	34	34	28	30	29	38	31	29	34	32	31
27	37	29	26	24	34	36	31	34	36	21	

We want to estimate the average speed for all cars in this area, based on the sample. Enter the data in a list. Here, I have entered them into list L1. This is a small sample – there are only 23 observations, so we should check to see if the data looks approximately normal.

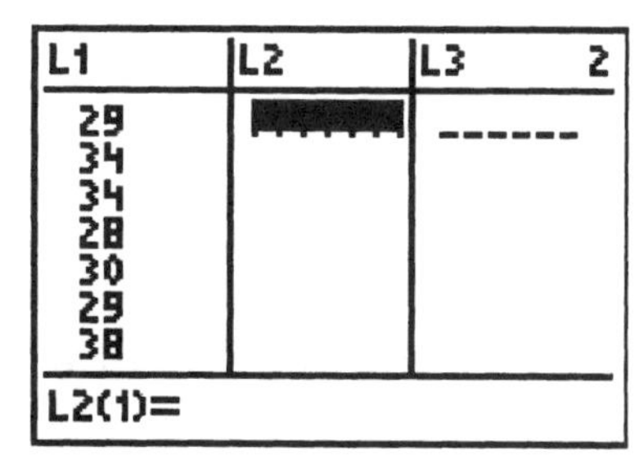

A normal plot of the data looks relatively straight, with no outliers, so it's reasonable to continue. This plot shows some granularity (repeated measurements of the same value) but no overt skewness. If you've forgotten how to create normal probability plots, return to chapter 4.

Press [STAT], arrow to TESTS then select choice 8:TInterval. You have two choices for data input: using data in a list such as we have or inputting summary statistics from the sample. Move the cursor to DATA and press [ENTER] to move the highlight. Enter the name of the list with the data ([2nd][1] for L1). Each observation occurred once, so leave Freq as 1. If there were a separate list of frequencies for each data value, that would be entered here. Enter the desired amount of confidence (here, 90%, but in decimal form) and finally press [ENTER] to perform the calculation.

Here are the results. Based on this sample, we are 95% confident the average speed for all cars on this road is between 29.5 and 32.6 miles per hour. There are two caveats here: the first is that this was not a truly random sample but a convenience one (only one 15 minute period was sampled). Also, the presence of the radar speed detector may have influenced the drivers at that time. Drivers may be driving over the posted 30 miles per hour limit, but since 30 is included in the interval, we have not shown that is wrong.

```
TInterval
 (29.523,32.564)
 x̄=31.04347826
 Sx=4.247761559
 n=23
```

What if we don't have the data? In the case of a small sample size, one must assume the data comes from an approximately normal population. If the sample is "large" the Central Limit Theorem will apply.

A nutrition laboratory tests 40 "reduced sodium" hot dogs, finding that the mean sodium content is 310 mg with a standard deviation of 36 mg. What is a 99% confidence interval for the mean sodium content of this brand of hot dog? Here we have moved the highlight from `Data` to `Stats`. When this is done, the input screen changes to ask for the sample mean, standard deviation, sample size and confidence level.

```
TInterval
 Inpt:Data Stats
 x̄:310
 Sx:36
 n:40
 C-Level:.99
 Calculate
```

Pressing ENTER to calculate the interval tells us we are 99% confident, based on this sample the mean sodium content for these hot dogs is between 294.6 and 325.4 mg.

```
TInterval
 (294.59,325.41)
 x̄=310
 Sx=36
 n=40
```

A one sample test for a mean

We can also do a hypothesis test to decide whether the mean speed is more than 30 mph. From the `STAT TESTS` menu, select choice `2:Ttest`. We are still using data in list L1. μ_0 is set to 30 since that's the speed limit we're comparing against. The alternate has been selected as $> \mu_0$ since we want to know if people are going too fast, on average. Notice we have the options of `Calculate` and `Draw` here, just as we did on tests of proportions.

```
T-Test
 Inpt:Data Stats
 μ0:30
 List:L1
 Freq:1
 μ:≠μ0 <μ0 >μ0
 Calculate Draw
```

Selecting `Draw` yields the screen at right. We can clearly see the shaded portion of the curve which corresponds to the p-value for the test of 0.1257. The calculated test statistic is $t = 1.178$. The p-value indicates we'll expect to see a sample mean of 31.04 (the mean from our sample) or higher by chance about 12.5% of the time by randomness when the mean really is 30. That's not very rare. We fail to reject the null and conclude these data do not show motorists on the street are speeding, on average.

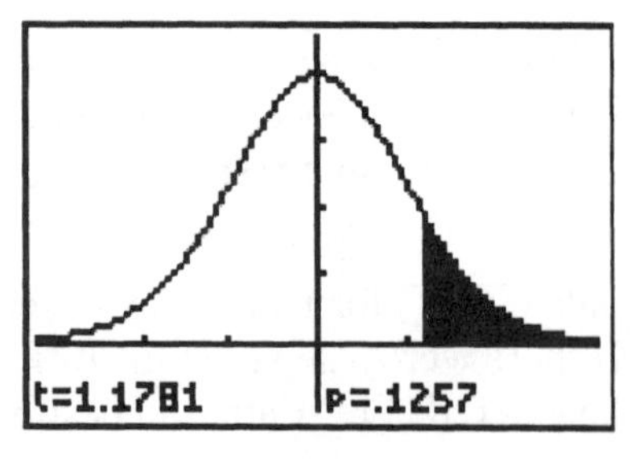

Comparing two means – confidence intervals

Should you buy name brand or generic batteries? Generics cost less, but if they do not last as long on average as the name brand spending the extra money may be worthwhile. Data were collected for six sets of batteries, which were used continuously in a CD player until no more music was heard through the headphones. The lifetimes (in minutes) for the six sets were:

Brand Name:	194.0	205.5	199.2	172.4	184.0	169.5
Generic:	190.7	203.5	203.5	206.5	222.5	209.4

The first step in performing a comparison such as this one (or any!) should always be to plot the data. Here, a side-by-side boxplot is natural.

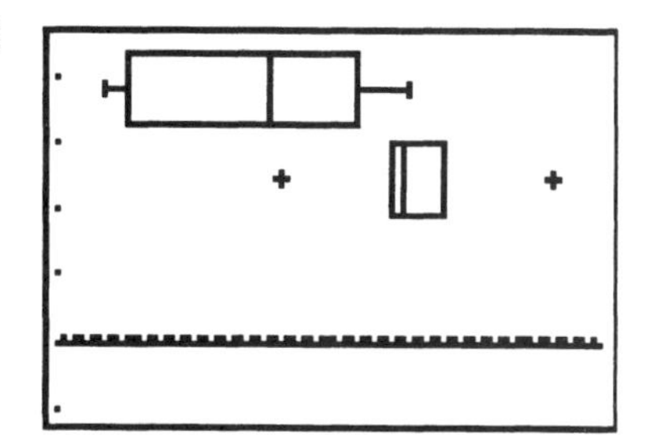

We have entered the data into list L1 for the Brand name batteries, and list L2 for the generics. We defined two boxplots to identify outliers on the STAT PLOT menu ([2nd][Y=]). For more on these plots, see chapter 3. From the plot, it certainly appears the generics last longer than the name brand batteries; they also seem more consistent (smaller spread). There are two outliers for the generic batteries, but with a sample size this small the outlier criteria are not very reliable. Neither of the extreme values are unreasonable, so it's safe to continue.

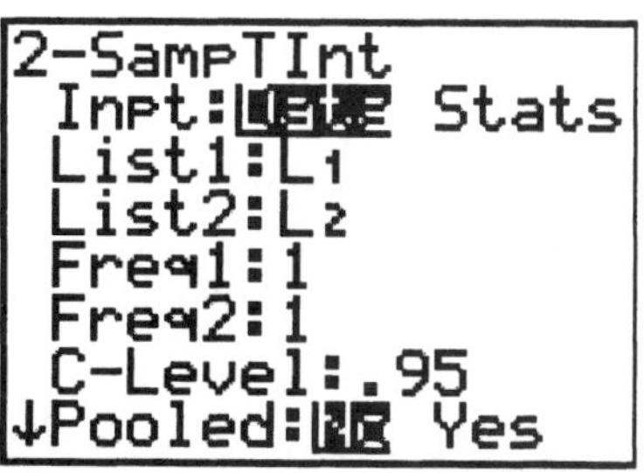

From the STAT TESTS menu select choice 0:2-SampTInt. Our data are already entered, so move the highlight (if necessary) to Data and press [ENTER]. The data were in lists L1 and L2, and each value in the lists occurred once. The confidence level has been set to 95% (entered as always in decimal form). The next option is new. Pooled: refers to whether the two groups are believed to have the same standard deviation or not. Visually, this is not true for our two battery samples. In general, unless there is some reason to believe the groups have the same spread, it's safest to answer this question with No. Reasoning behind this question has to do with computing a "pooled standard deviation" (or not) and the number of degrees of freedom for the test. Before the advent of computers (and statistical calculators) there were many recipes for handling this question, since the calculation of degrees of freedom in the unpooled case is complex. Luckily, we just let the calculator do the work.

```
2-SampTInt
 (-35.1,-2.069)
 df=8.986279467
 x̄1=187.4333333
 x̄2=206.0166667
 Sx1=14.6107723
↓Sx2=10.3019254
```

Pressing [ENTER] to calculate the interval gives the screen at right. We see we are 95% confident the average life of the name brand batteries is between 35.1 and 2.1 minutes *less* than the average life of the generic batteries. (Remember, it's always *group*1 – *group*2 in the interval). The next line gives the degrees of freedom for the interval – notice they're not even integer-valued. We also see the two sample means and standard deviations. The ↓ at the bottom left indicates more output can be obtained (the sample sizes). Assuming generic batteries are cheaper than name brand ones, it certainly would make sense to buy them.

Testing the difference between two means

If you bought a used camera in good condition, would you the same amount to a friend as to a stranger? A Cornell University researcher wanted to know how friendship affects simple sales such as this.[1] One group of subjects was asked to imagine buying from a friend whom they expected to see again. Another group was asked to imagine buying from a stranger. Here are the prices offered.

Friend:	$275	300	260	300	255	275	290	300
Stranger:	$260	250	175	130	200	225	240	

Here are side-by-side boxplots of the data. There certainly looks to be a difference. Prices to buy from strangers seem lower and much more variable than the prices for buying from a friend.

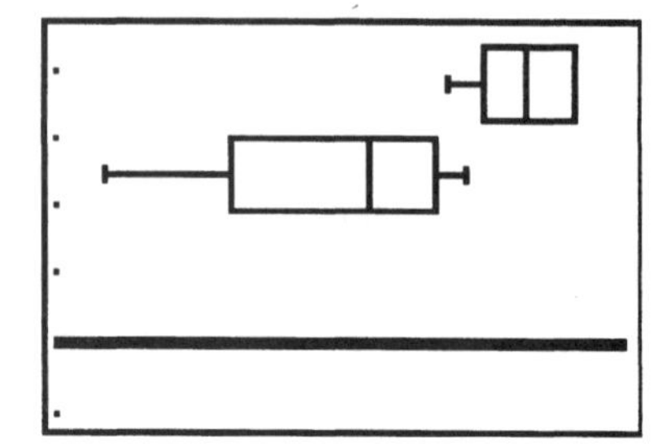

From the STAT TESTS menu, select choice 4:2-SampTTest. Again, we have the data in two lists, so Data is highlighted as the input mechanism, we have indicated the data are in lists L1 and L2, and each data value has a frequency of 1. The alternate hypotheses is $\mu_1 \neq \mu_2$ since our original question was "would you pay the same amount." Again, we have indicated No in regards to pooling the standard deviations (the spreads do not look equal and there is no reason to believe they should be the same).

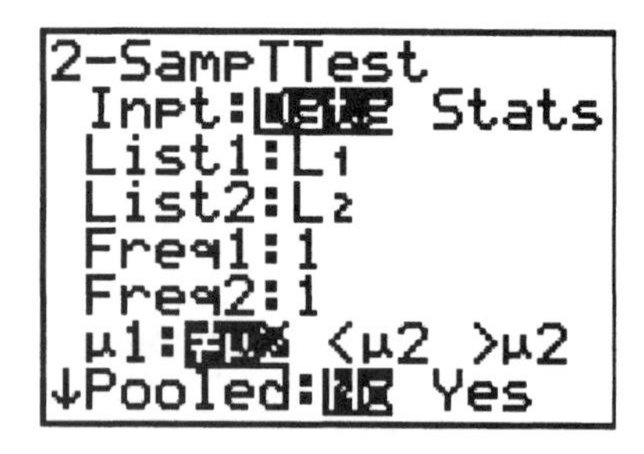

Pressing ENTER to calculate the test gives the screen at right. The computed test statistic is $t = 3.766$, and the p-value is 0.006. From these data we conclude that not only are people going to not pay the same amount to a friend than to a stranger, they're willing to pay more. We might even go so far as to warn people not to pay *too much* to friends.

2-SampTTest
μ1≠μ2
t=3.766049006
p=.0060025794
df=7.622947934
x̄1=281.875
↓x̄2=211.4285714

Paired Data

The two sample problems considered above used two *independent* samples. Many times data which might seem to be for two samples are naturally paired (say, examining the ages of married couples – each couple is a natural pair) or are even two observations on the same individuals. In such cases one works with the differences in each pair, and not the two sets of observations.

Do flexible schedules reduce the demand for resources? The Lake County (IL) Health Department experimented with a flexible four-day week. They recorded mileage driven by 11 field workers for a year

[1] Halpern, J.J. (1997). The transaction index: A method for standardizing comparisons of transaction characteristics across different contexts, *Group Decision and Negotiation*, 6(6), 557-572.

on an ordinary five-day week, then they recorded the mileage for a year on the four-day week.[2] Here are the data:

Name	5 day mileage	4 day mileage
Jeff	2798	2914
Betty	7724	6112
Roger	7505	6177
Tom	838	1102
Aimee	4592	3281
Greg	8107	4997
Larry G	1228	1695
Tad	8718	6606
Larry M	1097	1063
Leslie	8089	6392
Lee	3807	3362

Cursory examination reveals that after the change, some drove more, and some less. It is also easy to see there are large differences in the miles driven by the different workers. It is this variation between individuals that paired tests seek to eliminate.

We have entered the data into the calculator; the 5-day week mileages are in list L1, and the 4-day mileages are in list L2.

L1	L2	L3 3
2798	2914	
7724	6112	
7505	6177	
838	1102	
4592	3281	
8107	4997	
1228	1695	

L3(1)=

We need to find the differences. On the home screen, press [2nd][1][-][2nd][2][STO▶][2nd][3]. The command will look as at right. Pressing [ENTER] to complete the calculation will display the first few values. If you wish to see the entire list, go to the Statistics editor.

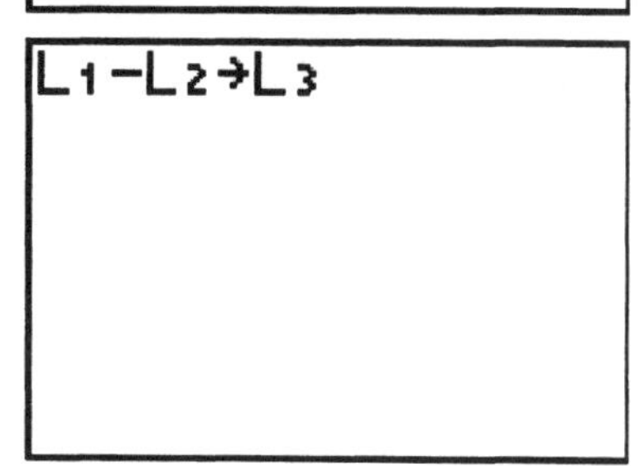

We need to check if the differences are approximately normal (or certainly at least no strong skewness or outliers). We define the normal plot as at right to use the differences which were just created. Press [ZOOM][9] to display the plot.

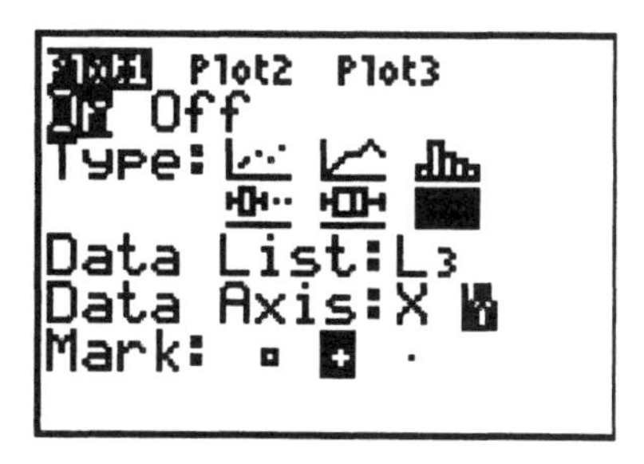

[2] Catlin, Charles S. Four-day Work Week Improves Environment, *Journal of Environmental Health*, Denver, March 1997 **59**:7.

The plot at right is not perfectly straight. However, there are no large gaps, so no extreme outliers.

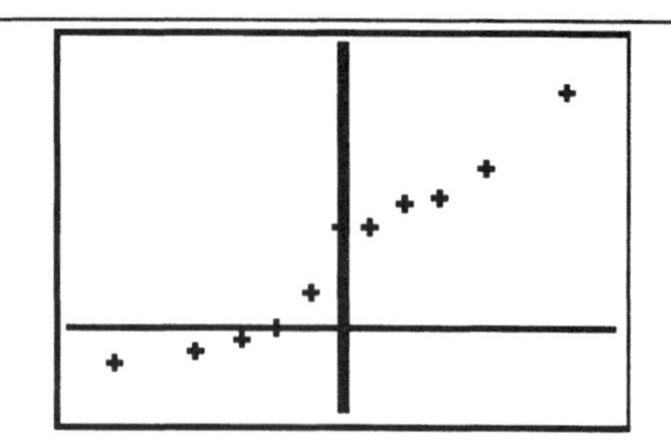

We now proceed to the test. We will perform a one-sample test using the differences as the data. From the STAT TESTS menu, select 2:T-Test. If the change in work week made no difference, the average value of the computed differences should be 0, so this is the value for μ_0. We are using the data from list L3 as the input, and have selected the alternative hypothesis as $\mu \neq \mu_0$.

```
T-Test
 Inpt:Data Stats
 μ0:0
 List:L3
 Freq:1
 μ:≠μ0 <μ0 >μ0
 Calculate Draw
```

Pressing ENTER when the cursor is over Calculate displays the results. The computed test statistic is $t = 2.85$ and the p-value is 0.017. We conclude that these data do indicate a difference in driving patterns between a 5-day work week and a 4-day work week. Further since the average difference is positive (982 miles) it seems that employees drove less on the 4-day week (the subtraction was 5-day – 4-day mileages). It's hard to say if the difference is meaningful to the department. If so, they may want to consider changing all employees to 4-day weeks.

```
T-Test
 μ≠0
 t=2.858034426
 p=.0170141282
 x̄=982
 Sx=1139.568339
 n=11
```

We can go further and compute a confidence interval for the average difference. Select 8:TInterval from the STAT TESTS menu, and define the interval as at right.

```
TInterval
 Inpt:Data Stats
 List:L3
 Freq:1
 C-Level:.95
 Calculate
```

Pressing ENTER to calculate the interval, we find we are 95% confident the 5-day work week will average between 216.4 and 1747.6 more yearly miles than a 4-day work week.

```
TInterval
 (216.43,1747.6)
 x̄=982
 Sx=1139.568339
 n=11
```

What can go wrong?

Not much that hasn't already been discussed – trying to subtract lists of differing length will give a dimension mismatch error. Having more plots "turned on" than are needed can also cause errors. The biggest thing to guard against is bad conclusions. Think about the data and what they show. Do not let conclusions contradict a decision to reject or not reject a null hypothesis.

Chapter 10 – Comparing Counts

Count data are analyzed primarily for three different purposes: whether or not data agree with a specified distribution (a goodness-of-fit test), whether or not observed distributions collected at different times or places are consistent with one another (a test of homogeneity), and whether or not data classified according to two categorical variables indicate the categorical variables are related or not (a test of independence). All of these tests use a probability distribution called the $\chi 2$ (chi-squared) distribution. The first test, that of goodness-of-fit, is not a built-in function on the TI-83, but can be done easily enough. The other two tests are built into the calculator and the mechanics are exactly the same. What is different is the setting and conclusions which can be made.

The $\chi 2$ statistic is defined to be $\sum \frac{(Obs - Exp)^2}{Exp}$ where the sum is taken over all the cells in a table. The quantities $\frac{(Obs - Exp)}{\sqrt{Exp}}$ are the standardized residuals which are examined in the event the null hypothesis is rejected to determine which cells deviated most from what is expected.

Testing Goodness-of-fit

Does your zodiac sign determine how successful you will be in later life? Fortune magazine collected the zodiac signs of 256 heads of the largest 400 companies. Here are the number of births for each sign.

Births	Sign
23	Aries
20	Taurus
18	Gemini
23	Cancer
20	Leo
19	Virgo
18	Libra
21	Scorpio
19	Saggitarius
22	Capricorn
24	Aquarius
29	Pisces

We can see some variation in the number of births per sign, but is it enough to claim that successful people are more likely to be born under certain star signs than others? If there is no difference between the signs, each should have (roughly) 1/12 of the births. That's the null hypothesis in this situation: births are evenly distributed across the year. The alternate is that the null is wrong: births are not evenly distributed across the year.

Here, the observed counts have been entered into L1 and the (hypothesized) probability for each birth sign (1/12) has been entered into L2. The calculator displays the decimal equivalent for the entered fraction.

L1	L2	L3 2
18	.08333	
21	.08333	
19	.08333	
22	.08333	
24	.08333	
29	.08333	

L2(13) =

Note: make sure both lists are the same length!

This next step calculates the quantities to be summed into the $\chi2$ statistic. The expected values for each cell are the proportions times the total number of individuals in the study (or 256/12 in this example). One needs to be careful in entering the command – the parentheses are necessary! Here the command was entered by pressing the following:
[(] [2nd] [1] [−] [2] [5] [6] [×] [2nd] [2] [)] [x^2] [÷] [(] [2] [5] [6] [×] [2nd] [2] [)] [STO▶] [2nd] [3].
After pressing [ENTER] the first few entries are displayed.

```
(L1-256*L2)²/(25
6*L2)→L3
{.1302083333 .0…
```

We need to add the entries in L3 to find the $\chi2$ value. Press [2nd] [STAT] (List), arrow to MATH then press [5] to select option 5 : sum(. The command shell is transferred to the home screen. Press [2nd] [3] for L3 followed by [ENTER]. The $\chi2$ statistic is 5.09375.

```
(L1-256*L2)²/(25
6*L2)→L3
{.1302083333 .0…
sum(L3
             5.09375
```

Now we need to get a p-value for the test. These tests are always one-tailed, so the p-value corresponds to area between 5.09375 and ∞ under a curve with n-1 degrees of freedom, where n is the number of categories (here there are 12 categories, so 11 degrees of freedom). Press [2nd] [VARS] (DISTR). We want choice 7 : χ2cdf(. Either arrow to the selection and press [ENTER] or press [7]. The command shell is transferred to the home screen.

```
DISTR DRAW
1:normalpdf(
2:normalcdf(
3:invNorm(
4:tpdf(
5:tcdf(
6:X²pdf(
7↓X²cdf(
```

The parameters are low end (5.09375), high end (properly [1] [2nd] [,] [9] [9] for infinity, but practically several 9's will work), then the degrees of freedom. Be sure to separate the parameters with a comma. The p-value for the test is 0.927 which is very large. We will not reject the null hypotheses and conclude that these data indicate no birth sign differences among the executives.

```
6*L2)→L3
{.1302083333 .0…
sum(L3
             5.09375
X²cdf(5.09375,99
9999,11)
        .9265413914
```

If the null were rejected, an examination of standardized residuals would show which cells were most different. The standardized residuals are just the square root of the entries in our list which had the components of $\chi2$. To find them, from the home screen press [2nd] [x^2] [2nd] [3] [)] [STO▶] [2nd] [4] followed by [ENTER]. To further examine these, use the Statistics list editor. We see (not surprisingly) the birth sign most different from its expected value is Pisces, the last entry in the list.

```
sum(L3
             5.09375
X²cdf(5.09375,99
9999,11)
        .9265413914
√(L3)→L4
{.3608439182 .2…
```

Tests of Homogeneity

Many high schools survey graduating classes to determine their plans for the future. We might wonder whether plans have stayed roughly the same or have changed through time. Here is a summary table from one high school. Each cell of the table shows how many students from each graduating class (the columns) had that particular type of plan (the rows). (Source IHS)

	1980	**1990**	**2000**
College/Post HS Educ	320	245	288
Employment	98	24	17
Military	18	19	5
Travel	17	2	5

Visually, choices do not appear to be the same (look at the row for Employment) but is the difference real or is it due perhaps to different size classes? Since we have really the same distribution at different time points, this is a test of homogeneity: the null hypothesis is that the distribution of students' plans is the same across time, the alternate hypothesis is that the distributions are not the same.

We will first enter the numbers in the body of the table into a matrix. Press [MATRX]. Arrow to EDIT. Press [ENTER] to select matrix A.

First we need to give the size. The body of our table had 4 rows and 3 columns, so the matrix is 4 x 3. Press [4][ENTER][3][ENTER] to change the size of the matrix. Now type in the entries in the body of the table following each by [ENTER]. The process goes left to right, top to bottom.

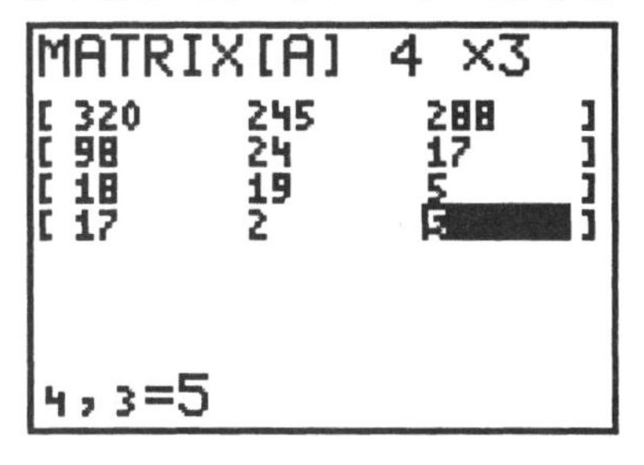

Here is the filled-in matrix. Press [2nd][MODE](QUIT) to leave the Matrix Editor.

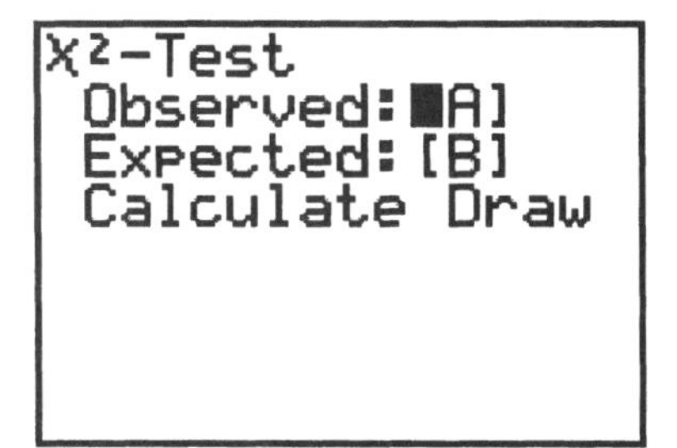

Now we're ready to perform the test. Press [STAT] then arrow to TESTS. We want choice C: χ2-Test. Either press [ALPHA][PRGM] (C) or press the up arrow to find the option followed by [ENTER]. All we need to tell the calculator is where the observed counts were entered and where to store expected counts. These default to matrix A for Observed and matrix B for Expected. If you need to change them, press [MATRX] then select the desired matrix. Notice we again have choices Calculate and Draw.

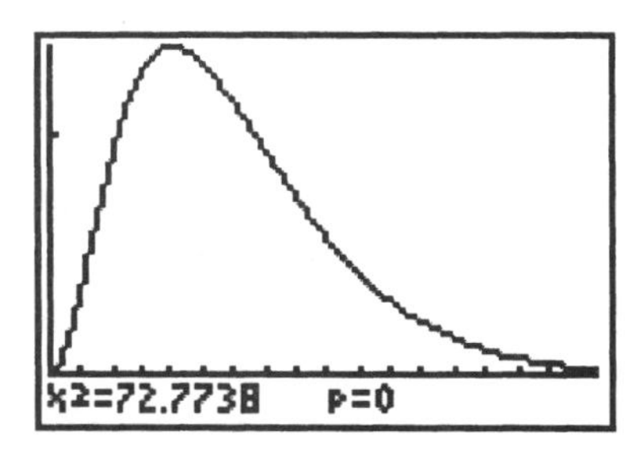

Selecting Draw gives the screen at right. We see three things here: the χ2 statistic (72.7738) and the p-value to four decimal places (0). We also see a different sort of distribution curve. χ2 curves are not symmetric; they are right skewed.

If we had chosen `Calculate` instead of `Draw` we would have this screen. It shows the same statistic value, but a little more exact p-value (but this is still essentially zero). We are also given the degrees of freedom for the test which are $(rows-1)(cols-1)$, so (4-1)(3-1) = 6.

```
X²-Test
 X²=72.77377814
 p=1.101778E-13
 df=6
```

Since the p-value is so small, we reject the null hypothesis and conclude the distributions of post-graduation plans are not the same for all three years.

Where are the differences? We'd like to get the standardized residuals. Unfortunately, the TI-83 won't give these easily. We can easily compute the matrix Obs-Exp. The Observed counts are in matrix A and the Expected counts are in matrix B. From the home screen, press [MATRX], press [ENTER] to select matrix A, then [–] press [MATRX] press the down arrow and select matrix B by pressing [ENTER] then press [STO▸][MATRX] and select a new matrix (probably C). Lastly, another [ENTER] will perform the calculation and display some of the results.

```
[A]-[B]→[C]
[[-45.22589792 ...
 [38.48487713  ...
 [.0170132325  ...
 [6.724007561  ...
```

Return to the matrix editor to see the entire matrix. The signs here indicate whether or not the observed count was more than the expected (positive) or less than expected (negative). Clearly the largest entries (in absolute value) are those for 1980 graduates with college plans and for 1980 graduates with plans for employment.

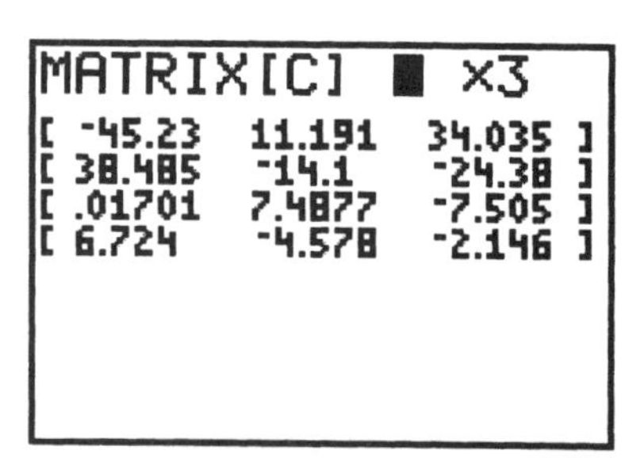

Here is matrix B which contains the expected counts. If we divide the entries in matrix C by the square root of the corresponding entry in matrix B we have the standardized residuals. The standardized residual 1980 graduates with college plans is $-45.23/\sqrt{365.23} = -2.37$. Similarly we find the standardized residual for 1980 graduates with employment plans is 4.98.

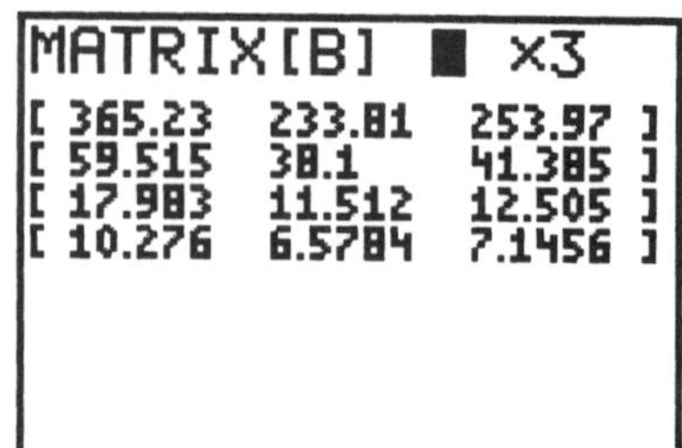

Testing Independence

Tests of independence are used when the same individuals are classified according to two categorical variables. A study from the University of Texas Southwestern Medical Center examined whether the risk of Hepatitis C was affected by whether people had tattoos and by where they got their tattoos. The data from this study can be summarized in a two-way table as follows.

	Hepatitis C	**No Hepatitis C**
Tattoo, Parlor	17	35
Tattoo, Elsewhere	8	53
No Tattoo	22	491

Is the chance of having hepatitis C independent of (not related to) tattoo status? Our null hypothesis is that the two are not related. The alternate hypothesis is that there is a relationship.

Enter the numbers from the body of the table into a 3 x 2 matrix as discussed above. Press [2nd][MODE] (Quit) to exit the matrix editor.

MATRIX[A] 3 ×2
[17 35]
[8 53]
[22 491]
3,2=491

The $\chi 2$ test is performed just as it is for the test of homogeneity. Here are the results. The p-value is extremely small. We will reject the null hypothesis and conclude there is a relationship between hepatitis C status and tattoos.

χ²-Test
χ²=57.91217384
p=2.657855E-13
df=2

Here are the results of computing Obs-Exp and storing the result into a new matrix as above. The largest standardized residual is $13.096/\sqrt{3.9042} = 6.628$. People with tattoos from tattoo parlors are more likely than normal to have hepatitis C. Perhaps tattoo parlors are a source of hepatitis C?

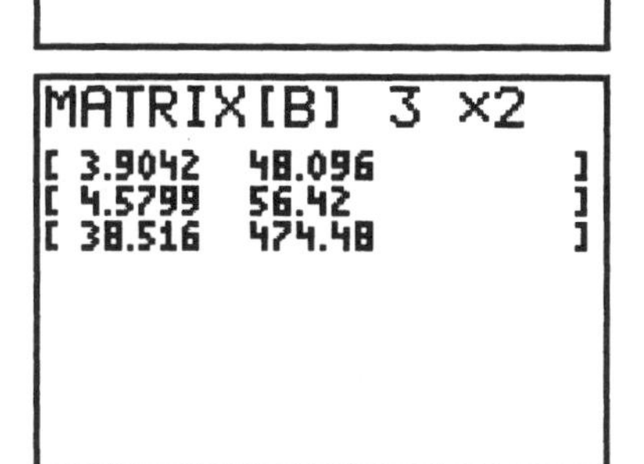

Here is the matrix of expected cell counts. Not all of them are more than 5. This means the conclusions based on the $\chi 2$ test we just performed may not be valid. Since the largest standardized residual is for one of these cells, this is a problem. A common solution is to combine cells in some manner to overcome the problem. In this case, both rows for tattooed people could be combined into one.

MATRIX[B] 3 ×2
[3.9042 48.096]
[4.5799 56.42]
[38.516 474.48]

What can go wrong?

Expected cell counts less than 5.

Check the computed matrix of expected cell counts. If they are not all greater than 5 the analysis may be invalid.

Missing or misplaced parentheses.

When computing elements for the goodness-of-fit test the parentheses are crucial.

Overusing the test.

These tests are so easy to do and data from surveys and such are commonly analyzed this way. The problem that arises here is that in this situation the temptation is to check many questions to see if relationships exist; but performing many tests on *dependent* data (the answers came from the same individuals) such as this is dangerous. In addition, remember that, just by random sampling, when dealing at $\alpha = 5\%$ we'll expect to see something "significant" 5% of the time when it really isn't. This danger is magnified when using repeated tests - it's called the problem of multiple comparisons.

Chapter 11 – Inference for Regression

Computing a regression equation and looking at residuals plots is not the end of the story. We might want to know if the slope (or correlation) is meaningfully different from 0. It's not always apparent that a slope is meaningfully non-zero. Consider these two equations for the selling price of a house: $price = 25 + 0.061 * sqft$ and $price = 25000 + 61 * sqft$. At first blush one might look at the small value for the slope in the first and believe it's reasonable to say the true slope may in fact be 0; however the difference is in the units – the first has price measured in thousands of dollars, the second in dollars. They're really the same line. In addition, we'd like to (perhaps) make a confidence interval for a "true" slope just as we did for means and proportions as well as confidence intervals for the average value of y for a given x and prediction intervals for a new y observation for an x value. The TI-83 calculator can perform the t-test on the slope as a native function. The other functions can be performed either using the calculator output and tables for the t-distribution or with a program which is included in this manual.

Returning to a problem considered before, here are advertised horsepower ratings and expected gas mileage for several 2001 vehicles.

Audi A4	170 hp	22 mpg	Buick LeSabre	205	20
Chevy Blazer	190	15	Chevy Prism	125	31
Ford Excursion	310	10	GMC Yukon	285	13
Honda Civic	127	29	Hyundai Elantra	140	25
Lexus 300	215	21	Lincoln LS	210	23
Mazda MPV	170	18	Olds Alero	140	23
Toyota Camry	194	21	VW Beetle	115	29

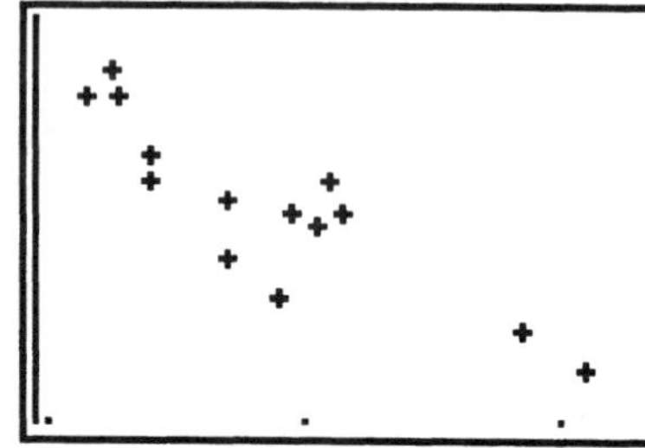

How is horsepower related to gas mileage? Recall the plot that was constructed for this data in Chapter 5. It is reproduced at right. The trend is decreasing. The residuals plots in Chapter 5 showed no overt pattern against X (horsepower) and the normal probability plot was reasonably straight. Inference for the regression is therefore appropriate.

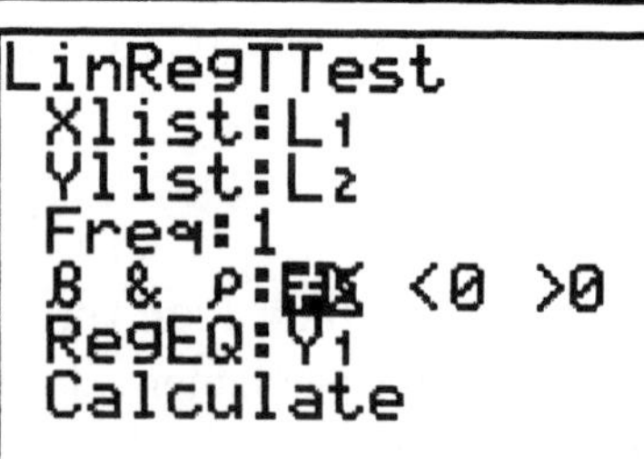

Press [STAT], arrow to TESTS and select choice E:LinRegTTest. You tell the calculator which list contains the x (predictor variable) values, which contains the y (response) values. Freq is normally set to 1. Indicate the appropriate form of the alternate hypothesis. Notice there is an option to store the equation of the line. To store the equation as a function (here, Y_1), press [VARS], arrow to Y-VARS, press [ENTER] to select Function, and [ENTER] to select Y_1. Finally, with the highlight on Calculate, press [ENTER].

This is the first portion of the output (notice the ↓ at the bottom left). The first lines indicate the form of the regression so that you are reminded which quantity is the slope (b) and which the intercept (a) and the form of the alternate hypothesis in the test. The computed t-statistic for this regression is –6.35 and the p-value for the test is 0.00004, with 12 degrees of freedom. We will reject the null hypothesis and conclude not only that the slope is not zero; it is significantly negative. The intercept for the regression is 38.45.

```
LinRegTTest
 y=a+bx
 β≠0 and ρ≠0
 t=-6.352711727
 p=3.64959E-5
 df=12
↓a=38.45416426
```

Pressing the down arrow several times we find the rest of the output. The slope is –0.092. We already know this is significantly different from zero even though its value seems small. The standard deviation of the data points around the line is 3.03. The relationship is strongly negative since $r^2 = 77.1\%$ and r = -0.877

```
LinRegTTest
 y=a+bx
 β≠0 and ρ≠0
↑b=-.0918175268
 s=3.032346154
 r²=.7708040509
 r=-.8779544697
```

A confidence interval for the slope

Confidence intervals (for any quantity) are always $estimate \pm (criticalvalue)(SE(estimate))$. In this case the critical value of interest will be a t statistic based on 12 degrees of freedom. From tables, we find this is 2.179 for 95% confidence. The standard error of the slope is

$$SE(b_1) = \frac{s(e)}{\sqrt{\sum (x-\bar{x})^2}} = \frac{s(e)}{\sqrt{n-1} * s(x)}.$$

```
1-Var Stats
 x̄=185.4285714
 Σx=2596
 Σx²=525390
 Sx=58.1889684
 σx=56.07229299
↓n=14
```

Using 1-Var Stats for our horsepower data in L1, we find Sx is 58.189.

We have all the pieces we need. $SE(b_1) = \dfrac{3.0323}{\sqrt{13} * 58.189} = 0.0145$.

When computing this, be sure to enclose the denominator in parentheses and close the parenthesis for the square root. Putting all the pieces together, the 95% confidence interval for the slope is $-0.092 \pm 2.179 * 0.0145$ or (-0.124, -0.060). Based on this regression, we are 95% confident average gas mileage decreases between –0.124 and –0.060 miles per gallon for each horsepower in the engine.

A confidence interval for the mean at some X

What should we predict as the average gas mileage for a vehicle with 160 horsepower? Evaluating the equation for 160 horsepower gives 23.76 miles per gallon. This is just a point estimate, however and is subject to uncertainty just as any mean is. Confidence intervals account for this uncertainty – in this case there are two sources – average variation around the line as well as uncertainty about the slope which makes estimation more "fuzzy" further away from the mean. Both of these are accounted for in the equation of the standard error, $SE(\hat{\mu}_v) = \sqrt{s^2(b_1) * (x_v - \bar{x})^2 + \dfrac{s^2(e)}{n}}$. Putting everything together, we find

$SE(\hat{\mu}_v) = \sqrt{0.0145^2 * (160 - 185.429)^2 + \dfrac{3.0323^2}{14}} = 0.890$. The t critical value is still 2.179, so the confidence interval is 23.76 ± 2.179*0.890 or (21.82, 25.70). Based on this regression, we estimate with 95% confidence the average gas mileage for vehicles with 160 horsepower will be between 21.82 and 25.70 miles per gallon.

A prediction interval for a new observation

What would we predict for gas mileage for a particular vehicle with 160 horsepower? The point estimate is still 23.76 miles per gallon, but we have some additional uncertainly because individual observations are more variable than means. The standard error becomes $SE(\hat{y}_v) = \sqrt{s^2(b_1)*(x_v - \bar{x})^2 + \frac{s^2(e)}{n} + s^2(e)}$ which becomes $SE(\hat{\mu}_v) = \sqrt{0.0145^2 * (160 - 185.429)^2 + \frac{3.0323^2}{14} + 3.0323^2} = 3.160$. So the prediction interval is 23.76 ± 2.179*3.160 or (16.88, 30.65). Based on this regression we estimate with 95% confidence the gas mileage for a vehicle with 160 horsepower will be between 16.88 and 30.65 miles per gallon.

**A program for Regression Inference

The author of this manual has written a program performs these functions. (A listing is included and the program can also be obtained from the Web site which can be downloaded into a TI–83). The program name is LSCINT. Once the program has been loaded into the calculator, to run the program, press [PRGM] and select that name from the list of programs. PgrmLSCINT is transferred to the home screen. Press [ENTER] to start the program. You are prompted for the X list, enter its name and press [ENTER]. You will then be prompted for the Y list; enter its name and press [ENTER].

```
prgmLSCINT
X LIST=L1
Y LIST=L2
```

The next screen gives the coefficients in the equation, the correlation coefficient (r) and the coefficient of determination, r^2 and well as the standard deviation of the residuals (s). Press [ENTER] to continue.

```
Y=a+bX
a=38.45416426
b=-.0918175268
r=-.8779544697
r²=.7708040509
S=3.032346154
```

These are the results of the t-test for the slope. Press [ENTER] to continue.

```
t=-6.352711727
df=12
p=3.649589965E-5
```

You will next be prompted for a confidence level (enter it as a decimal) and the value of X for which confidence and prediction intervals will be created. Press [ENTER] after inputting each value.

C LEVEL=.95
X=?160

The calculator displays the confidence interval for the slope. Press [ENTER] to continue.

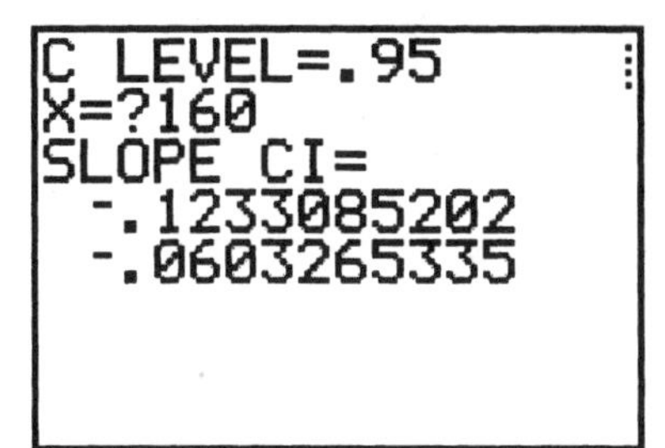

Now the calculator displays the y-value for the given x and confidence and prediction intervals. Press [ENTER] to finish the program.

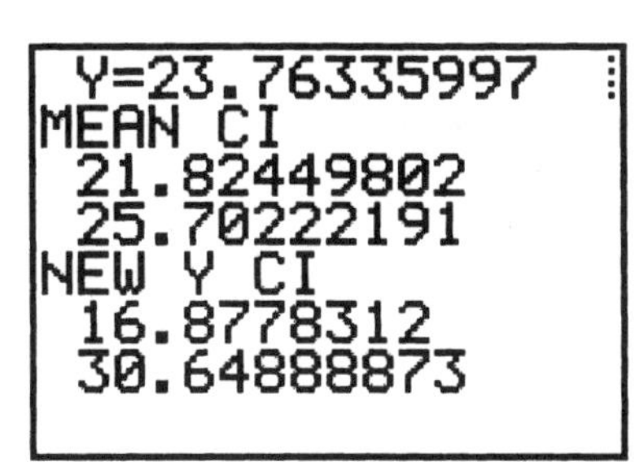

What can go wrong?

Assuming lists are the same length, not much that has not already been covered. One problem in doing many of these computations "by hand" comes from the compounding of round-off errors in intermediate computations. One is generally safest in using many digits in the interim and rounding only at the end. (Notice the "hand calculated" intervals are somewhat different from those obtained from the calculator. This is the reason.)

Program LSCINT listing (TI-83 Ascii version) This can also be found on the Web site.

```
Input "X LIST=",LX
Input "Y LIST=",LY
FnOff
LinRegTTest LX,LY,0,Y1
σx^2n[STO▶]V:s[STO▶]S:Ë[STO▶]M:n[STO▶]N
ClrHome
Output(1,2,"Y=a+bX"
Output(2,2,"a="
Output(2,4,a
Output(3,2,"b="
Output(3,4,b
Output(4,2,"r="
Output(4,4,r
Output(5,2,"r^2="
Output(5,5,r^2
Output(6,2,"S="
Output(6,4,S
Pause
df[STO▶]K
ClrHome
Output(1,1,"t="
Output(1,3,t
Output(2,1,"df="
Output(2,4,K
Output(3,1,"p="
Output(3,3,p
Pause
ClrHome
Input "C LEVEL=",C
Prompt X
K+1[STO▶]K:b[STO▶]A
TInterval 0, √ (K),K,C
upper[STO▶]T
A-T*S/√ (V)[STO▶]P
A+T*S/√ (V)[STO▶]Q
Output(3,1,"SLOPE CI="
Output(4,2,P
Output(5,2,Q
Pause
Y1-ST√(N^-1+(X-M)^2/V)[STO▶]P
Y1+ST√(N^-1+(X-M)^2/V)[STO▶]Q
Y1-ST√(1+N^-1+(X-M)^2/V)[STO▶]J
Y1+ST√(1+N^-1+(X-M)^2/V)[STO▶]U
ClrHome
```

```
Output(1,2,"Y="
Output(1,4,Y1
Output(2,1,"MEAN CI"
Output(3,2,P
Output(4,2,Q
Output(5,1,"NEW Y CI"
Output(6,2,J
Output(7,2,U
Pause
ClrHome
```

Chapter 12 – Analysis of Variance (ANOVA)

We have already seen two-sample tests for equality of the means in Chapter 9. What if there are more than two groups? Answering the question relies on comparing variation among the groups to variation within the groups, hence the name. The null hypothesis for ANOVA is always that all groups have the same mean and the alternate is that at least one group has a mean different from the others.

Wild irises are beautiful flowers found throughout North America and northern Europe. Sir R. A. Fisher collected data on the sepal lengths in centimeters from random samples of three species. The data are below. Do these data indicate the mean sepal lengths are similar or different?

Iris setosa	Iris versicolor	Iris virginica
5.4	5.5	6.3
4.9	6.5	5.8
5.0	6.3	4.9
5.4	4.9	7.2
5.8	6.7	6.4
5.7	5.5	5.7
4.4	6.1	
	5.2	

I have entered the data into lists L1, L2, and L3. We will first construct side-by-side boxplots of the data for visual comparison. Visually, the medians are somewhat different with Iris virginica being the largest.

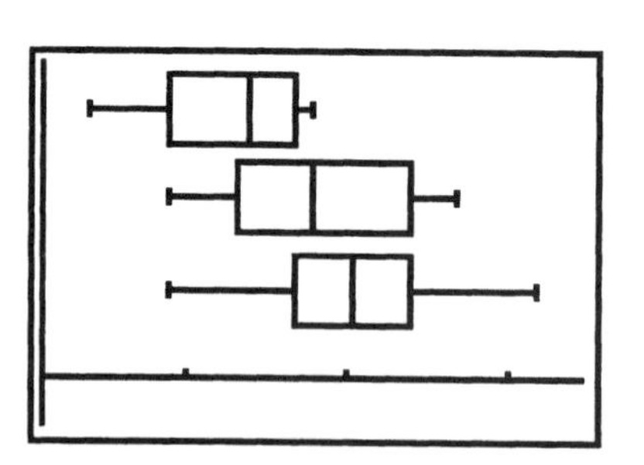

From the STAT TESTS menu select choice F:ANOVA(. This is the last test on the menu, so it is easiest to find by pressing the up arrow. The command shell is transferred to the home screen.

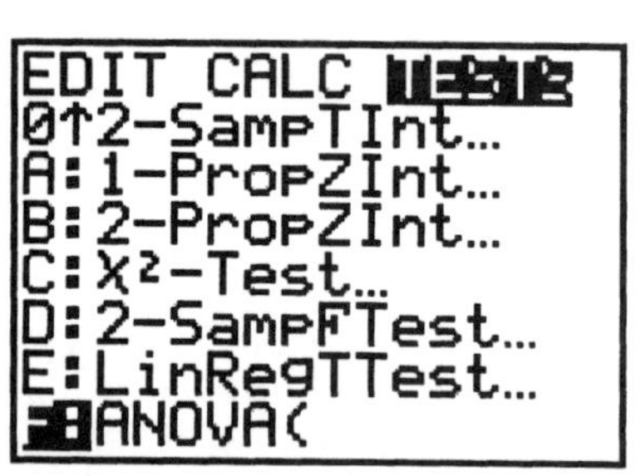

To complete the command, enter the list names separated by commas, then press [ENTER].

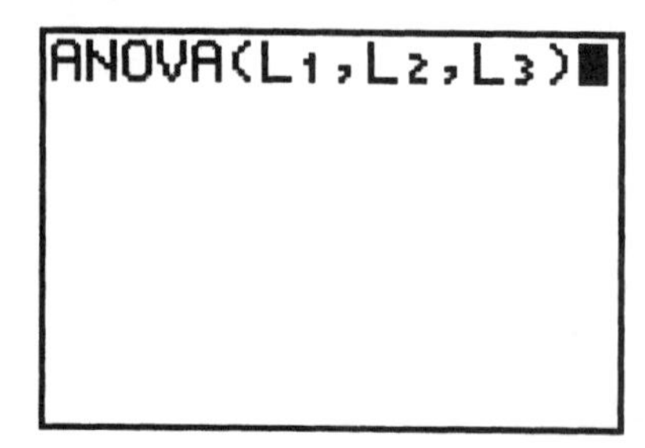

This is the first portion of the output. The value of the F statistic is 2.95 and the p-value for the test is 0.0779 which indicates at $\alpha = 5\%$ there is not a significant difference in the mean sepal lengths for the three species, based on this sample. The Factor degrees of freedom are k-1 where k is the number of groups, so with three groups, this is 2. MS is SS/df.

```
One-way ANOVA
 F=2.95022657
 p=.0779466278
 Factor
  df=2
  SS=2.4414881
↓ MS=1.22074405
```

Pressing the down arrow several times gives the remainder of the output. Degrees of freedom for Error are n-k, where n is the total number of observations in all groups (21 here) and k is the number of groups (3). MS is again SS/df. Sxp is the estimate of the common standard deviation and is the square root of MSE. The F statistic is MSTR/MSE.

```
One-way ANOVA
↑ MS=1.22074405
 Error
  df=18
  SS=7.44803571
  MS=.413779762
 Sxp=.643257151
```

Another Example

The following data represent yield (in bushels) for plots of a given size under three different fertilizer treatments. Does it appear the type of fertilizer makes a difference in mean yield?

Type A	Type B	Type C
21	41	35
24	44	37
31	38	33
42	37	46
38	42	42
31	48	38
36	39	37
34	32	30

Here are side-by-side boxplots of the data. All three distributions appear symmetric and there is not a large difference in spread, so ANOVA is appropriate. Using 1-Var Stats we find the mean for Type A is 32.125 bushels, Type B has a mean of 40.125 bushels, and Type 3 has a mean of 37.25. They're different, but are they different enough?

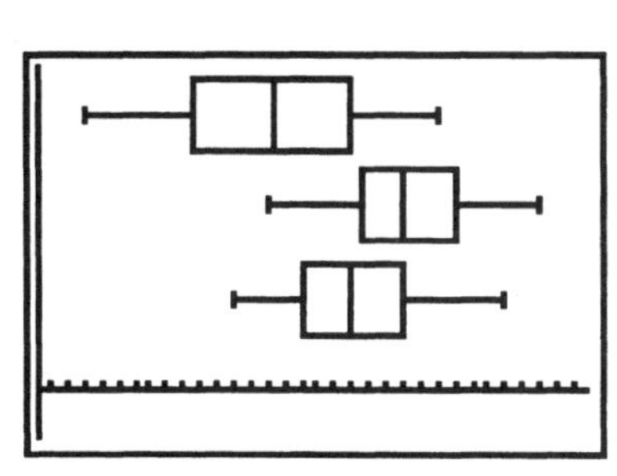

The ANOVA output of interest is at right. We see the F-statistic is 4.05 and the p-value for the test is 0.0325. At $\alpha = 0.05$, we will reject the null hypothesis and conclude that at least one fertilizer has a different mean yield than the others. Scrolling down, we further find Sxp = 5.696.

```
One-way ANOVA
 F=4.049724771
 p=.0325471129
 Factor
  df=2
  SS=262.75
↓ MS=131.375
```

Which mean(s) is (are) different?

Having rejected the null, we would like to know which mean (or means) are different from the rest. For a rough idea of the difference one can do confidence intervals for each mean (using the pooled standard deviation found in the ANOVA) and look for intervals which do not overlap, but this method is flawed. Similarly, testing each pair of means (doing three tests here) has the same problem: the problem of multiple comparisons. If we constructed three individual confidence intervals, the probability they all contain the true value is $0.95^3 = 0.857$, using the fact that the samples are independent of each other.

The Tukey method is a way to solve this question. From tables, we find the critical value q^* for k groups and n-k degrees of freedom for error. My table gives $q^*_{3,21,0.05} = 3.58$ The "honestly significant difference" is computed as $HSD = \frac{q^*}{\sqrt{2}} s_p \sqrt{\frac{2}{n_i}}$ where each group has n_i observations. (NOTE: most tables of this distribution require the division of q^*, but not all. Check your particular table.) Here, $HSD = \frac{3.58}{\sqrt{2}} * 5.695\sqrt{\frac{2}{8}} = 7.208$. Groups with means that differ by more than 7.208 are significantly different. The mean for Type A was 32.125; for Type B 40.125 and for Type C 37.25. We therefore declare types A and C have similar means; Types B and C also have similar means. The means which are different from each other are those for types A and B.

Chapter 13 – Multiple Regression

Multiple regression is an extension of the linear regression already studied where we create a model to explain a response variable based on more than one predictor. Just as with linear regression, we will want to examine how well the predictors individually and as a group determine the response by testing the utility of the model and create confidence intervals for slopes, mean response, and predictions of new responses.

How well do age and mileage determine the value of a used Corvette? The author chose a random sample of ten used Corvettes advertised on autos.msn.com. The data are below.

Age (Years)	Miles (1000s)	Price ($1000)
3	46	27
1	11	43
2	20	35.5
1	11.5	39
8	69	16.5
5	49	23
2	10	38
4	27	32
5	30.5	30
3	46	27

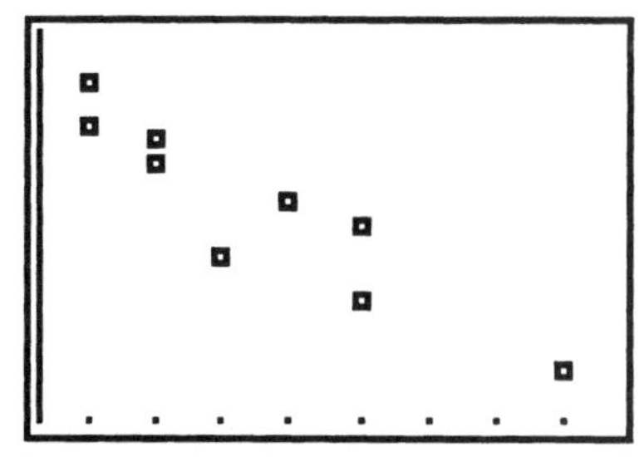

We first examine plots of each predictor variable against Price. The plot against age is linear, and decreasing as expected (we expect older cars to cost less). The regression equation for this relationship is *Price* = 42.44 – 3.33**Age*, with $r^2 = 80.6\%$. This suggests the average price of a used Corvette goes down $3330 each year.

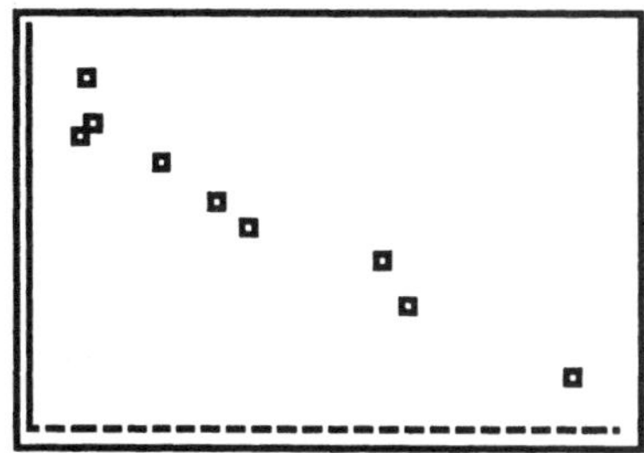

The plot of Price against mileage is also linear and decreasing with even less scatter than in the other plot. The regression equation for this relationship was found to be *Price* = 43.77 – 0.40**Miles*, with $r^2 = 95.5\%$. This suggests the average price of a used Corvette will decrease $400 for every 1000 miles it has been driven.

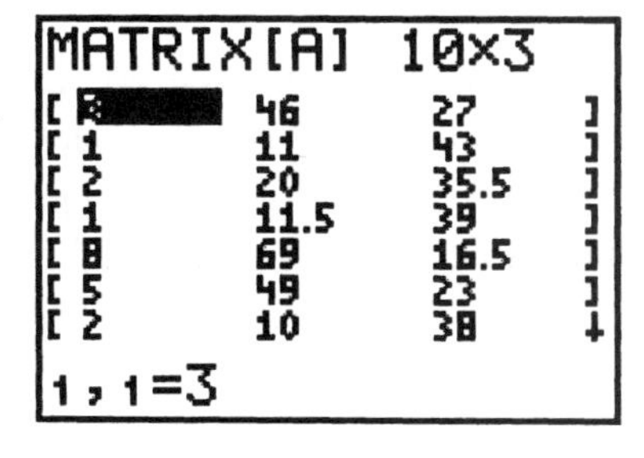

The TI-83 cannot do multiple regression on its own. The author has written a program (provided on the web site) called MULREG to do this. First the data are entered into Matrix A. Press [MATRX], arrow to EDIT and press [ENTER] to select matrix A. Enter the number of observations (rows, here 10) and the number of columns (3). Type in the data, pressing [ENTER] after each number, across the rows of the matrix. Press [2nd][MODE] (Quit) to exit the editor.

With the program transferred to the calculator, pressing [PRGM] will give the list of all programs stored. Select MULREG, then press [ENTER] to start it running. You will first be asked which column of the matrix has the response (Y) variable. In our example, Price is in the third column. Press [ENTER] after the response.

```
                 Done
prgmMULREG
DATA IN COLS
OF [A]
RESPONSE COL=3
```

Here is the first portion of the output, the coefficients. These are displayed with the program paused so you can use the right arrow to scroll through them. They are also stored in list $\theta 1$ which can be accessed under the LIST menu ([2nd][STAT]). Press [ENTER] to resume execution of the program. We find the equation of the model is *Price* = 44.200 – 0.943**Age* – 0.309**Kmiles*.

```
                 Done
prgmMULREG
DATA IN COLS
OF [A]
RESPONSE COL=3
COEF Lθ1=
{44.2002058 -.9…
```

The next quantities displayed are the standard deviations of the coefficient estimates. As with the coefficients themselves, the calculator is paused to allow scrolling through the list. These are stored in list $\theta 2$ for further use. After pressing [ENTER] the T statistics for each coefficient are displayed in a like manner, followed by the P-values for testing the hypotheses H_0: $\beta_i = 0$ against H_A: $\beta_i \neq 0$.

```
RESPONSE COL=3
COEF Lθ1=
{44.2002058 -.9…
STDEV Lθ2=
{.9457508901 .4…
T-RATIO Lθ3=
{46.7355688 -2.…
```

The next screen first displays S, the standard deviation of the residuals, then R^2, the coefficient of determination which is how much of the variation in response (Price) is explained by the model, 97.4%. The next quantity shown is the adjusted R^2. Since R^2 can never decrease when additional variables are added into a model, this quantity is "penalized" for additional variables which do not significantly help explain variation in Y, so it will go down if this is the case. Adjusted R^2 is always less than the regular R^2. We also see the degrees of freedom which are associated with the F-test for overall utility of the model. Press [ENTER] to resume execution of the program.

```
S=          1.487591
R²=         .9735386
R²ADJ=      .9659782
REG DF= 2
ERR DF= 7
TOT DF= 9
```

This screen shows the sums of squares and the F statistic for the overall significance (utility) of the model, along with its p-value. Here the p-value (to 5 decimal places) is 0 indicates there is a significant relationship between price and its two predictors.

```
SS REG= 569.9095
SS ERR= 15.49049
SS TOT= 585.4
MS REG= 284.9547
MS ERR= 2.212928
F=      128.7681
P-VAL=  0
```

Now we are asked if we want to use the model to predict a value based on the equation, or quit. Make the appropriate choice.

Assessing the model

Just as with simple (one-variable) linear regression, we will use residuals plots to assess the model. The program has stored fitted values (y-hats) in L5 and residuals in L6.

Using the STAT PLOT menu, define a scatter plot of the residuals in L6 (as Y) against the fitted values in L5. Pressing [ZOOM][9] displays the plot. Just as with simple regression we are looking in this plot for indications of curves or thickening/narrowing which indicate problems with the model. With this small data set these are somewhat hard to see, but clearly the only positive residuals are for the largest and smallest fitted values, which could indicate a potential problem.

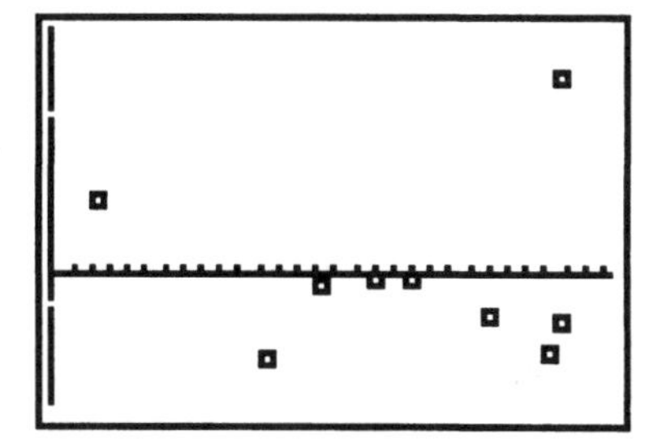

We define a normal plot of the residuals (as in Chapter 4). This normal plot is not a straight line, which indicates a violation of the assumptions. This multiple regression model is not appropriate for these data.

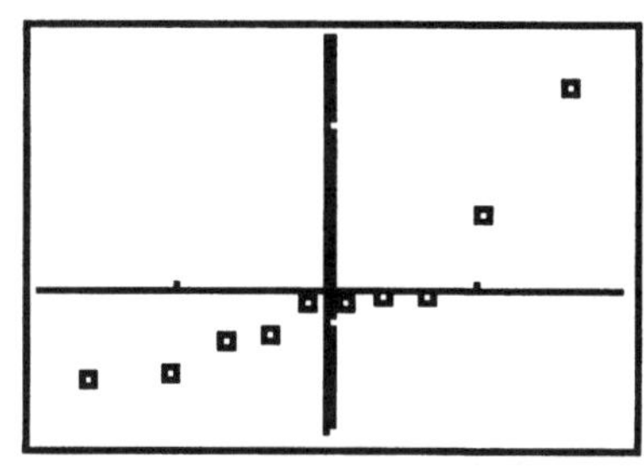

Multiple Regression Confidence Intervals

How well do the midterm grade and number of missed classes predict final grades? Data for a sample of students are below.

Final Grade, Y	Midterm Exam, x_1	Classes Missed, x_2
81	74	1
90	80	0
86	91	2
76	80	3
51	62	6
75	90	4
48	60	7
81	82	2
94	88	0
93	96	1

Performing the regression as above, we find the estimated regression equation $Final = 49.41 + 0.502 * Midterm - 4.71 * ClassesMissed$. All coefficients are significantly different from zero and a normal plot of the residuals is relatively straight. We'd like to use the model to create a prediction of the average grade for students with a midterm grade of 75 and 2 absences (a confidence interval); we also want to predict the grade for a particular student with a midterm grade of 75 and 2 absences (a prediction interval).

When executing the MULREG program, select 1:YES in answer to the question PREDICT Y. You will be asked to input the independent variables' values. Enter then inside curly braces ([2nd][{] and [2nd][}] separated by commas. You will then be asked for the desired confidence level.

```
X{}={75,2}
CONF. LEVEL=.95
```

Pressing [ENTER] performs the calculations and displays the screen at right. We find the point estimate of the final grade for a student with a 75 midterm grade and two absences is 77.66 or 78 (practically speaking). The standard deviation of the fit at that point is 0.392. The confidence interval says we are 95% confident the average final grade for all students with a 75 midterm grade and 2 absences will be between 76.7 and 78.6. Further, we are 95% confident an individual student with a 75 midterm grade and 2 absences will earn a final grade between 75.3 and 80. Pressing [ENTER] at this point returns you to the Predict or Quit menu.

```
Y HAT= 77.66251
S(FIT)=.39171
C.I.=
  76.73626494
  78.58875506
P.I.=
  75.32188955
  80.00313045
```

What can go wrong?

Not much that hasn't already been discussed. The most common errors are misspecification of lists and having more than one plot "turned on" at a time.

How do I get rid of those extra lists?

Press [2nd][+] (**MEM**). Select `2:Delete`, then select `4:List`. Arrow to the lists to be deleted and press [ENTER].

MULREG Program Listing. (The program can also be downloaded from the Web site.)

```
Disp "DATA IN COLS","OF [A]"
dim([A]) STO▶θ1:Lθ1(1)STO▶N:Lθ1(2) STO▶L:L-1STO▶K
{N,1}STO▶dim([B]):Fill(1,[B])
augment([B],[A]) STO▶ [B]
[B]^T[B] STO▶ [D]
seq([D](I,1),I,2,L+1)/NSTO▶Lθ1
{L,L}STO▶dim([C]):AnsSTO▶dim([E])
{N,1}STO▶dim([C])
Input "RESPONSE COL=",R
For(I,1,N):[A](I,R) STO▶ [C](I,1):End
If R≤K:Then
For(J,R+1,L):For(I,1,N)
[B](I,J+1) STO▶ [B](I,J)
End:End
[B]^T[B] STO▶ [D]
End
{L,L}STO▶dim([D]):[D]x⁻¹STO▶ [D]
{N,L}STO▶dim([B])
[B]^T [C]:[D]AnsSTO▶ [E]
Matr▶list([E],Lθ1)
Disp "COEF Lθ1=":Pause Lθ1
[B][E] STO▶ [B]
Matr▶list([C],LY)
Matr▶list([B],LYP)
mean(LY) STO▶Y
sum((LY-Y)²)STO▶T
sum((LY-LYP) ²STO▶E
LYPSTO▶L5
LY-LYPSTO▶L6
DelVar LY:DelVar LYP
N-LSTO▶M:T-ESTO▶R:R/KSTO▶Q:E/MSTO▶D:√(D) STO▶S:Q/DSTO▶F
S√(seq([D](I,I),I,1,L)) STO▶Lθ2
Disp "STDEV Lθ2=":Pause Lθ2
Lθ1/Lθ2STO▶Lθ3
Disp "T-RATIO Lθ3=":Pause Lθ3
Disp "COEF P Lθ4="
1-2seq(tcdf(0,abs(Lθ3(I)),M),I,1,L) STO▶Lθ4
Pause Lθ4
1-(N-1)*D/TSTO▶A
ClrHome
Output(1,1,"S="
```

```
Output(1,9,S
Output(2,1,"R²="
Output(2,9,R/T
Output(3,1,"R²ADJ="
Output(3,9,A
Output(4,1,"REG DF="
Output(4,9,K
Output(5,1,"ERR DF="
Output(5,9,M
Output(6,1,"TOT DF="
Output(6,9,N-1
Pause
ClrHome
Output(1,1,"SS REG="
Output(1,9,R
Output(2,1,"SS ERR="
Output(2,9,E
Output(3,1,"SS TOT="
Output(3,9,T
Output(4,1,"MS REG="
Output(4,9,Q
Output(5,1,"MS ERR="
Output(5,9,D
Output(6,1,"F=    "
Output(6,9,F
Output(7,1,"P-VAL="
1-Fcdf(0,F,K,M) [STO▸]P
round(P,5) [STO▸]P
Output(7,9,P
Pause
ClrHome
Lbl C
Menu("PREDICT Y?","YES",A,"QUIT",B)
Lbl A
Input "X{}=",Lθ5
Input "CONF. LEVEL=",C
{L,1}[STO▸]dim([C])
1[STO▸] [C](1,1)
For(I,1,K):Lθ5(I) [STO▸] [C](I+1,1):End
[E]ᵀ[C]
Ans(1,1) [STO▸]Y
ClrHome
round(Y,5) [STO▸]Y
Output(1,1,"Y HAT="
Output(1,8,Y
[C]ᵀ[D][C]
Ans(1,1) [STO▸]V
```

```
round(S√(V),5) STO▸ T
Output(2,1,"S(FIT)="
Output(2,8,T
TInterval 0,√(M+1),M+1,C
upper STO▸ T
Output(3,1,"C.I.="
Y+TS√(V) STO▸ D
Y-TS√(V) STO▸ E
Output(4,3,E
Output(5,3,D
Output(6,1,"P.I.="
Y-TS√(1+V) STO▸ E
Y+TS√(1+V) STO▸ D
Output(7,3,E
Output(8,3,D
Pause
Goto C
Lbl B
DelVar [C]:DelVar [D]:DelVar [E]:DelVar Lθ5
ClrHome
```

Chapter 14 - TI-89 Basics

Figure 1.1 The TI-89 Calculator

This calculator has built-in symbolic manipulation capabilities. Although a few statistical functions are "native" on the TI-89, most of the topics covered in an Intro Stats course require downloading the Statistics with List Editor application which is free. Download requires the TI-Connect cable. See the web page http://education.ti.com/us/product/tech/89/apps/appslist.html for more information. This manual assumes the statistics application has been loaded on the calculator.

Commonly Used Buttons and Commands

[ON]	At the lower left, this button turns the calculator on. When first turned on, the calculator displays the "home" screen.
[2nd]	The gold key on the upper left. This changes the cursor to ▮ and is used to access items listed in gold above keys
[2nd][ON]	Turns the calculator off. It will automatically turn itself off to save battery life after a few minutes of inactivity.
[ENTER]	The key on the lower right. This is used to execute commands
[2nd] [ENTER]	Used to recall the last command entered. Used repeatedly, this cycles through the last several commands.
[CLEAR]	This key is on the upper right, just below the down arrow. It is used to clear commands on the home screen and lists in the Statistics Editor
[2nd][ESC]	Executes the quit command, which leaves a menu or exits the Statistics and Matrix editors.
[−]	Located on the right side, just above the + key. This is the arithmetic "minus."

[(-)] Located next to [ENTER]. This is the "negative" key. Don't confuse it with the [-].

[alpha] This is the purple key located near the upper left. This toggles the cursor between letters (indicated in purple above keys) and standard key functions.

[STO▸] Located just above [ON], this allows storage of many results.

[APPS] On the TI-89, this allows access to built-in applications as well as Flash applications which include Inferential Statistics.

[◆] The green key near the upper left, allows access to commands and functions in labeled in green above keys.

Cursor movement and Command Editing

⊖, ⊖ Up and down arrow keys are used to move the cursor around menus

◀, ▶ Left and right arrow keys will move the cursor one keystroke when editing commands. In graphics displays, with activated, they move the cursor around the plot

[2nd]◀, [2nd]▶ Move the cursor to the beginning or end of a command.

[◆][←] Located next to the left arrow key near the upper right. This deletes the character to the right of the cursor in a command.

[2nd][←] This is the Insert command which allows characters to be inserted when editing.

Graphics commands are labeled in green above the blue keys on the very top of the calculator. They are accessed by pressing [◆] followed by the appropriate choice of [F1] through [F5].

[Y=] A blue key located on the upper left. It is used to enter functions.

[WINDOW] Sets graphing windows; this is usually only necessary for histogram plots

[GRAPH] Displays a graph with the current window settings.

Adjusting the Display Contrast:

[◆][+]	(Hold down the [+] key) will make the screen darker.
[◆][−]	(Hold down the [−] key) will make the screen lighter.

Menu Operation

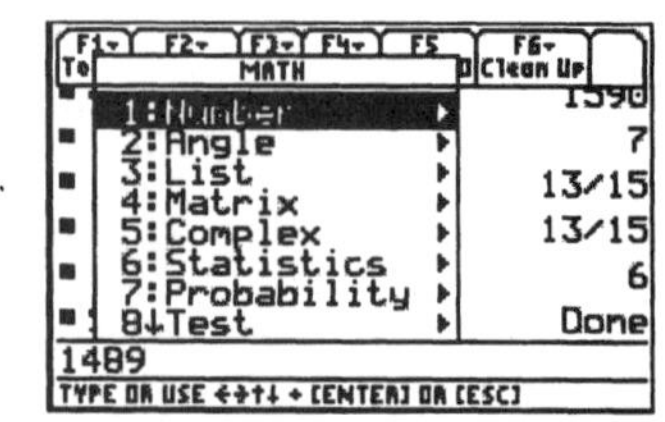

Many keys on the calculator activate menus which also may have sub-menus. For example, pressing [2nd][5] activates the MATH menu. There are numerous sub menus indicated by the ▶. To select a submenu option either press the [▼] or [▲] button until the desired choice is highlighted, then pressing [ENTER], or simply type the number (or alpha character) of the menu option. To display the submenu, press [▶]. Alternatively, simply press the number of the submenu.

Sharing Data and Programs between Calculators

Data and programs may be shared between calculators using the communications cable which is supplied. The TI-89 can only communicate with the other TI-89s and TI-92s.

Connect the supplied cable to the port at the base of each calculator. On both calculators press [2nd][−] to activate the Var-Link menu.

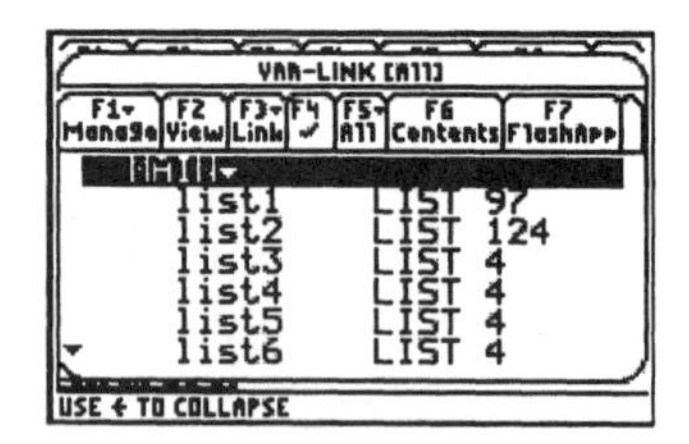

On the receiving calculator, press [F3] to select Link. [▼] to highlight RECEIVE, then press [ENTER]. The screen reverts to main Var-link Menu with a "`Waiting to receive`" message at the bottom.

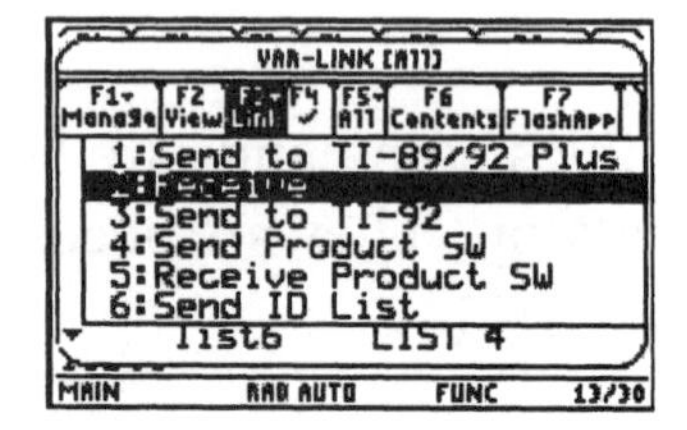

On the sending calculator, use the arrow keys to select the item to send, then press [F4] the "check" the item. The example at right will send list1 and list 2.

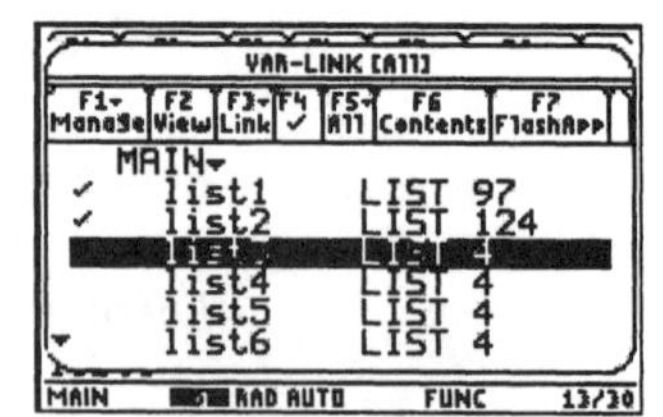

Now press [F3] to select Link, then [ENTER] to select menu choice 1:Send to TI-89/92Plus which is highlighted by default.

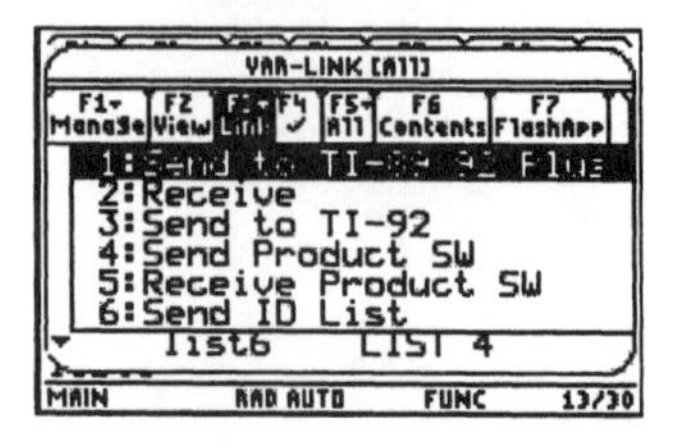

An analogous procedure can be used to send applications between calculators. Applications (such as the Statistics with List Editor) are selected from the F7 `FlashApp` menu (press [2nd][F2] for F7.)

Sharing Data between the Calculator and a Computer

Data lists, screen shots and programs may be shared between the calculator and either Microsoft Windows or Macintosh computers using a special cable and either TI-Connect or TI-Graph Link software. Cables are available for either serial or ISBN ports and can be found through many outlets such as OfficeMax, and Amazon.com. The software can be downloaded free through the Texas Instruments website at education.ti.com.

Working with Lists

The basic building blocks of any statistical analysis are lists of data. Before doing any statistics plot or analysis the data must be entered into the calculator. There are six basic lists available in the statistics editor (the TI-89 can actually have more through naming lists and using folders, but we won't go into that here) they are called list1 through list6. Accessing lists by name is done by pressing [2nd][−] (Var-LINK).

Accessing the Statistics Editor

Press [◆][APPS] followed by selection of the Flash Application name followed by [ENTER]. If this is the first time the editor has been accessed since the calculator has been turned on, you will be prompted for a data folder as in the middle screen. The default folder is main. Press [ENTER] to select main as the current folder, or press [▶] to allow a new folder to be created. To enter a new folder name, arrow to the entry block and type the name of the new folder. Your lists will be stored in the new folder and it will be set to default. To change folders, press [▶] to select a folder. Then press [ENTER] to proceed to the list editor. If the editor has been used since the calculator has been turned on, pressing [ENTER] to select the application will automatically open the editor.

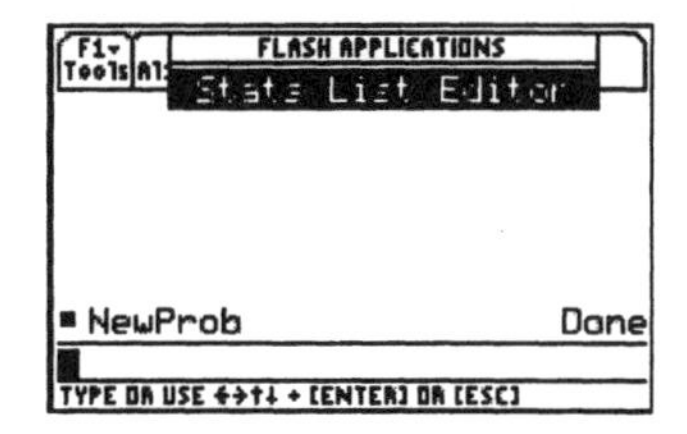

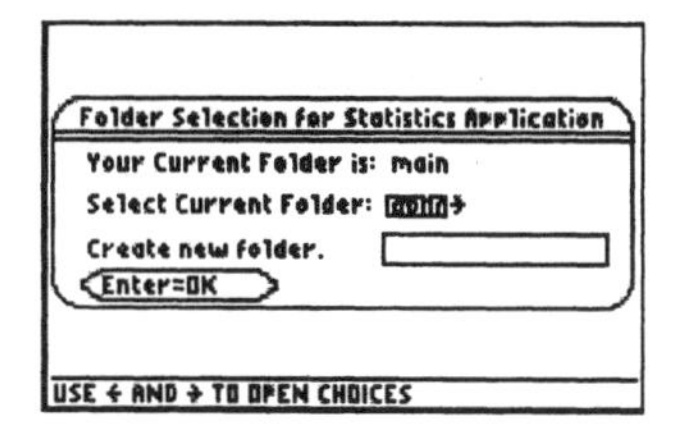

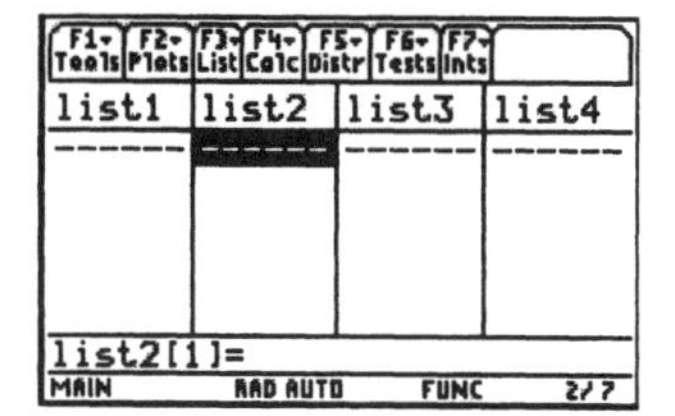

To enter data, simply use the right or left arrows to select a list, then type the entries in the list, following each value with [ENTER]. Note that it's not necessary to type any trailing zeros. They won't even be seen unless decimal places (found with the [MODE] key) have been set to some fixed number.

One word of advice: Most lists of data in texts are entered across the page in order to save space. Don't think that just because there are 4 (or more!) columns of data they belong in 4 (or more) lists. Data which belongs to a single variable always belongs in a single list.

Editing Data

It is always recommended that a list be double-checked for accuracy after entering it. Use the ⊖ and ⊖ keys to scroll through the list. If an entry is found to be in error, simply type the correction over the current value, then press [ENTER].

If a value needs to be inserted in the list, one can scroll to the bottom of the list and add it in the case of a single variable. However, if data are paired (such as in regression, frequency tables or paired tests) you will need to either delete the corresponding entry from the second list and place both at the bottom of each list or insert an item. Place the cursor on the value below where a new one is to be entered. Press [2nd][←] and a new line with a 0 value will be inserted. Move the cursor to the new value and type it in. In the example below, a new entry for the third element of the list is desired.

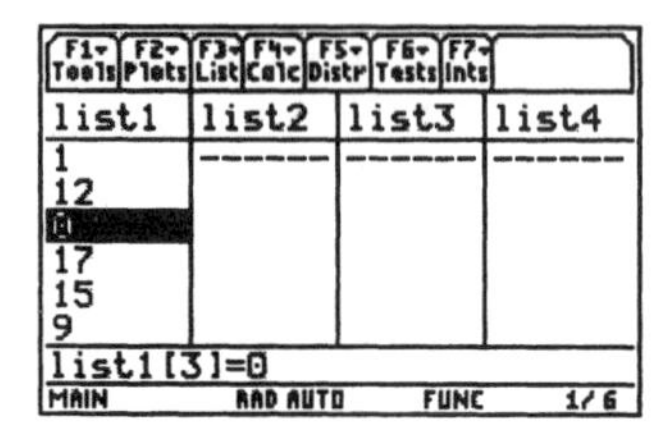

To delete an item, move the cursor to the desired item, then press [◆][←]. In the example below, the fourth element in list1 (17) is deleted.

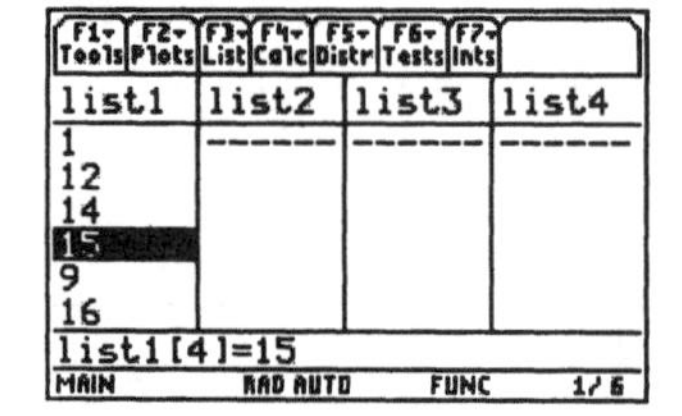

Exiting the Editor

To exit (leave) the statistics editor press [2nd][ESC] for quit. Note that it is not always necessary to leave the statistics editor before performing a new function. For example, if once a list is entered descriptive statistics are desired, press [F4] (CALC), press [ENTER] to select 1-Var Stats, then input the list name ([2nd][-] for VAR-LINK then select the appropriate list name) press [ENTER] to perform the calculation.)

Erasing Lists

There are several ways to erase (clear) all values in a list. Probably the easiest is to move the cursor so the name of the list is highlighted, then press [CLEAR][ENTER]. Never press [DEL] instead of [CLEAR].

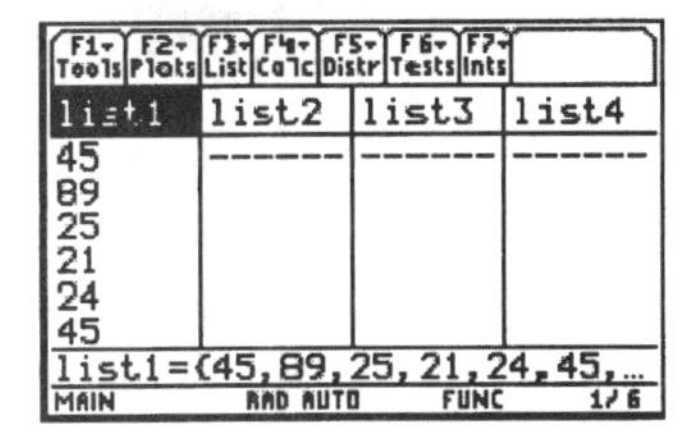

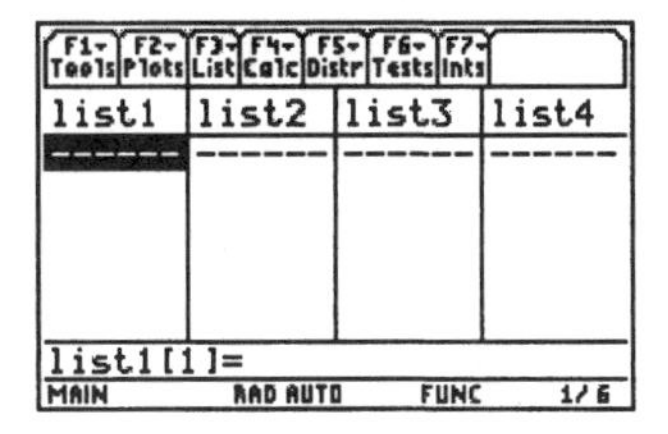

To erase several lists at once, from the home screen, press [CATALOG][F3] (Flash Apps), then arrow down to `clrList(` and press [ENTER]. The command is transferred to the command portion of home screen. Press [2nd][-] for the Var-Link menu, select lists to clear by arrowing to them and pressing [ENTER] to select them. Separate list names by commas. The list of list names must be finished by a right parenthesis. Press [ENTER] to execute the command. The example below clears list1.

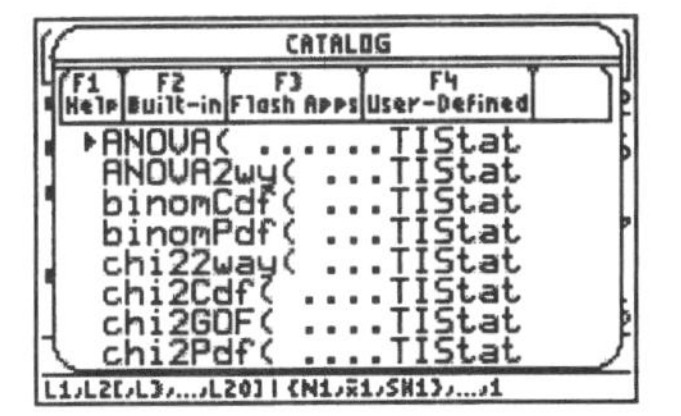

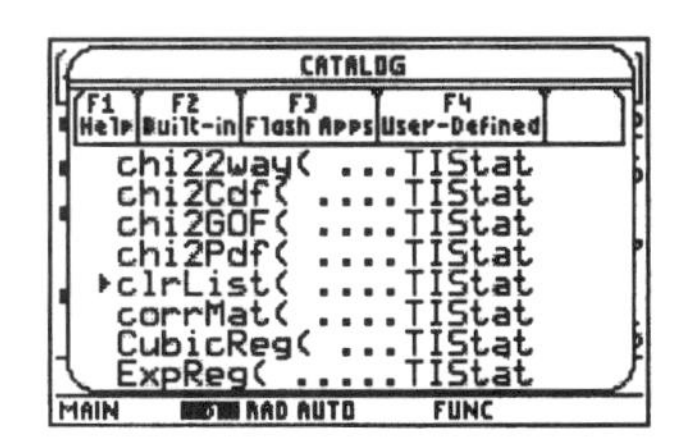

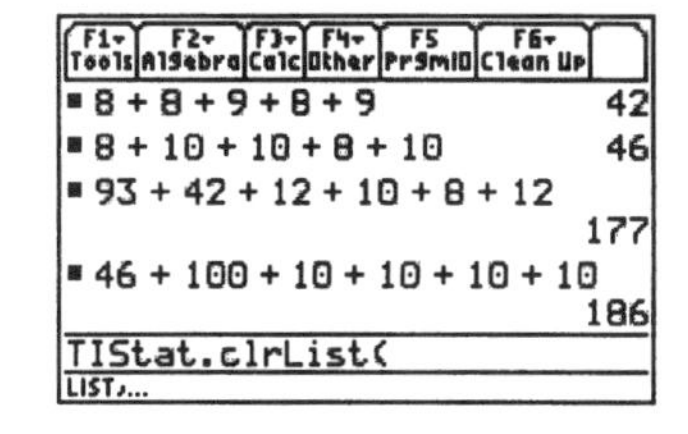

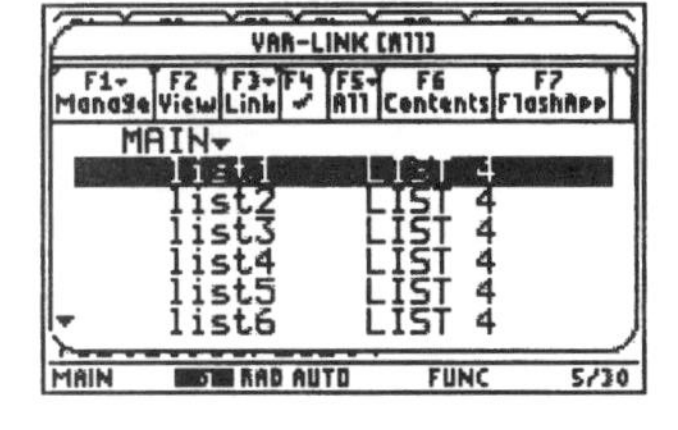

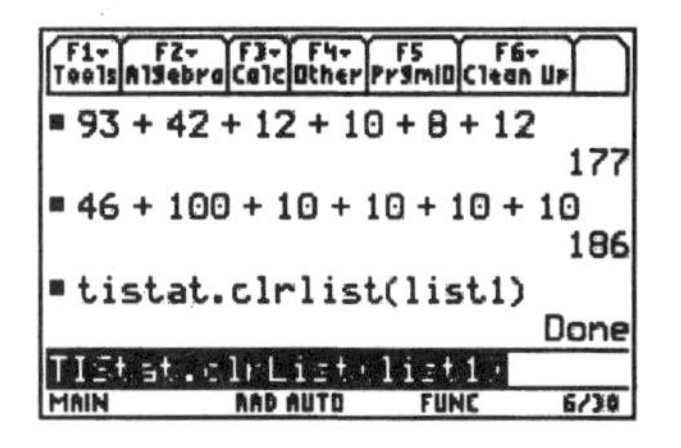

Sorting Lists

While in the List Editor, press F3 for the List menu, press [▼] or 2 to select `List Ops`, then press [ENTER] to Select 1:Sort List, [2nd][-] (`Var-Link`) to and use the arrows to select the list to be sorted. Arrow if necessary to change the sort order (use [▶] to activate the menu choices). Press [ENTER] to carry out the command.

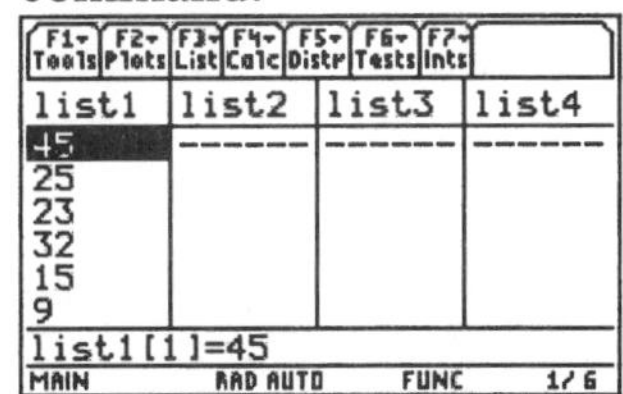

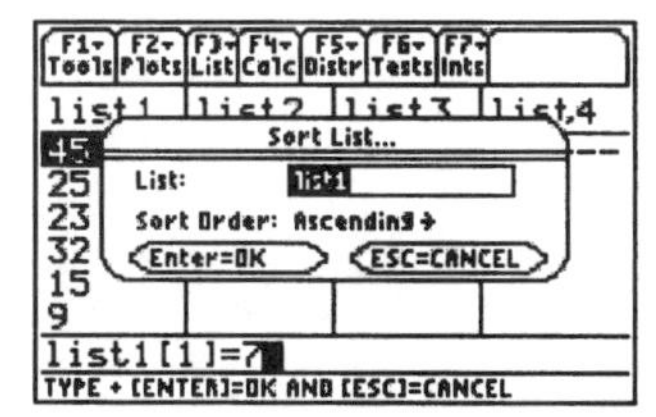

What Can Go Wrong?

Why is my list missing?
By far the most common error, aside from typographical errors is improper erasure of lists. When lists seem to be "missing" the user has used delete rather than [CLEAR] in attempting to erase a list. Believe it or not, the data and the list are still in memory. To reclaim the missing list, press [F1] (Tools) then select 3:Setup Editor. Leave the box named Lists To View blank and press [ENTER]. Lists 1 through 6 will now be available, and any extra lists will be erased.

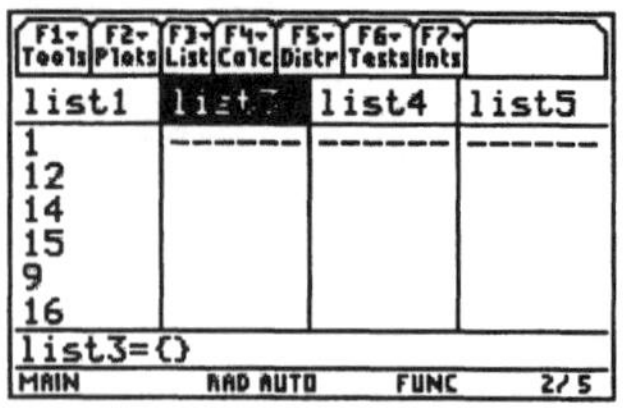

Chapter 15 – Displaying Quantitative Data

The three primary rules of data analysis are:

1. Make a picture.
2. Make a picture.
3. Make a picture.

TI-calculators can help with this, although they cannot make bar graphs or pie charts for categorical data or dotplots and stem-and-leaf displays. All of these are fairly easily done by hand (at least for small data sets; for larger ones, use a full computer package).

This chapter will discuss histograms and time plots. Other plots of quantitative data will be discussed later. Histograms are examined for shape (skewed right/left or symmetric), center, and spread. They also tell us whether or not a distribution in unimodal (one-humped) or multi-modal (many humps). Time plots give indications of trends (systematic rising or falling patterns) and cycles (repeating patterns of rise and fall) in data which are observations on the same variable at different time points; they can also be used to examine the nature of volatility (how short or tall the peaks and valleys are).

In this chapter we will primarily work with the following data which represent monthly stock price changes in dollars for Enron stock for the period January 1997 to December 2001, just before the company collapsed. Looking at a table with lots of numbers is not a good way to understand what they show and to see patterns in the data.

	Jan	**Feb**	**Mar**	**Apr**	**May**	**Jun**	**Jul**	**Aug**	**Sep**	**Oct**	**Nov**	**Dec**
1997	-$1.44	-0.75	-0.69	-0.88	0.12	0.75	0.81	-1.75	0.69	-0.22	-0.16	0.34
1998	0.78	0.62	2.44	-0.28	2.22	-0.50	2.06	-0.88	-4.50	4.12	1.16	-0.50
1999	3.28	3.34	-1.22	0.47	5.62	-1.59	4.31	1.47	-0.72	-0.38	-3.25	0.03
2000	5.72	21.06	4.50	4.56	-1.25	-1.19	-3.12	8.00	9.31	1.12	-3.19	-17.75
2001	14.38	-1.08	-10.11	-12.11	5.84	-9.37	-4.74	-2.69	-10.61	-5.85	-17.16	-11.59

Histograms

Histograms are connected barcharts. Since the data are presumed to represent particular observations on some (underlying) continuous portion of the real number line and since order here matters, bars are always displayed connected to one another (unless there happens to be a gap in the values.) A good histogram has equal bar widths, high and low ends not too dramatically different from the maximum and minimum values of the data, and intervals which "make sense." As we will see, they are useful in showing major features of the distribution of a single variable or for comparing two distributions (if done properly); they also have a capacity for "artiness" since their shape can change, depending on the choice of beginning values and bar widths.

The first step in making a histogram is to enter the data. The Enron data have been entered into `list1`; the first few values are seen in the accompanying figure. Notice that it's not necessary to type all the leading 0's when doing the data entry. Also, be sure to lead the negatives with the [(-)] key not the [-] (minus) key.

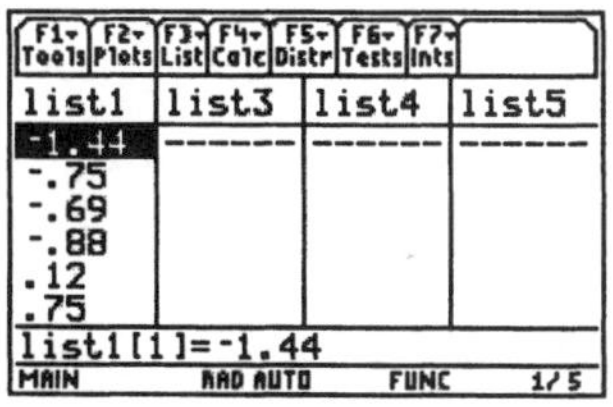

Steps to Create a Histogram

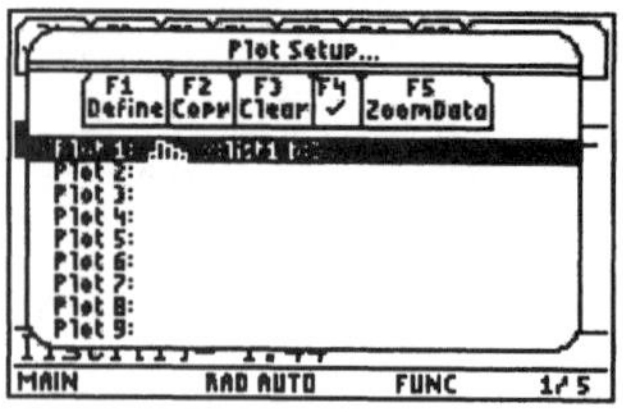

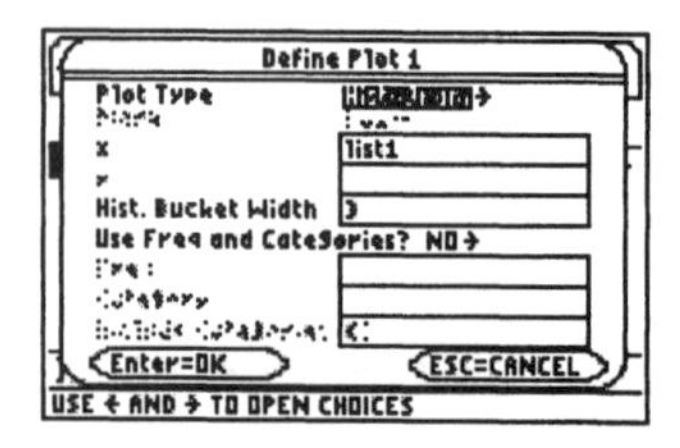

The next step is to define the plot. This is done from the Statistics Editor by pressing [F2] (Plots) followed by [ENTER] to select Plot Setup. You will see the screen at right. Notice that there are nine plots which can be displayed at any one time. For most purposes, there should be only one active (checked) plot at once.

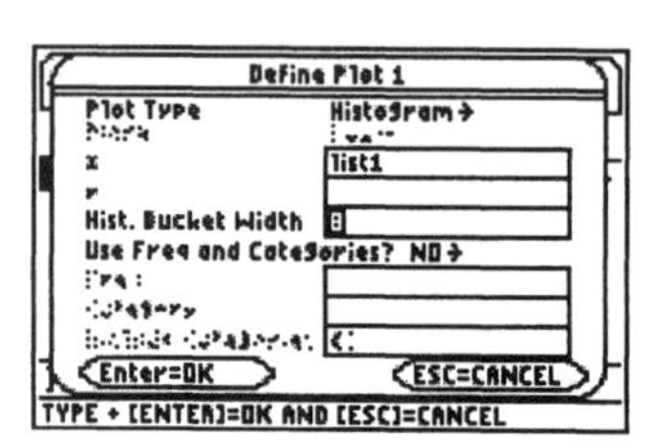

Press [F1] to select defining Plot1. The cursor should be blinking over the plot type. If not already set to a histogram, pressing the right arrow gives a menu of five plot types. Move the cursor to highlight choice 4:Histogram and press [ENTER] to select it or press [4].

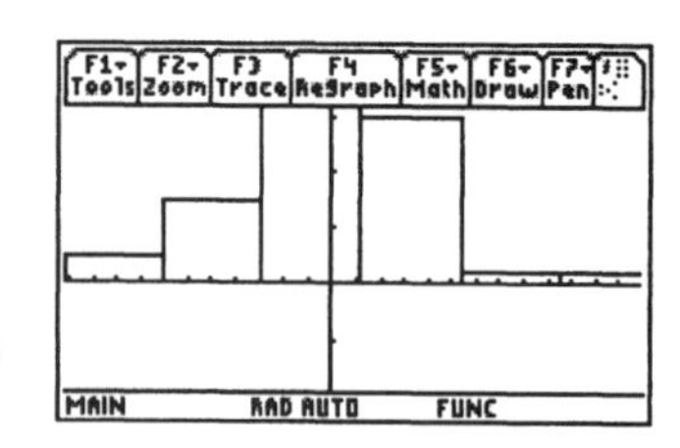

Press the down arrow to the box labeled x. Press [2nd][-] (VAR-LINK) to get the list of list names. Move the cursor to highlight the one you wish to use, then press [ENTER] to select it. The TI-89 then wants the histogram bucket width which is the bar width. Press the down arrow to move to this box. This is something that may have to be "played around with" to get a good picture. Here, I have set the bar width to 8. Press [ENTER] to complete the plot definition. You will be returned to the Plot Setup menu.

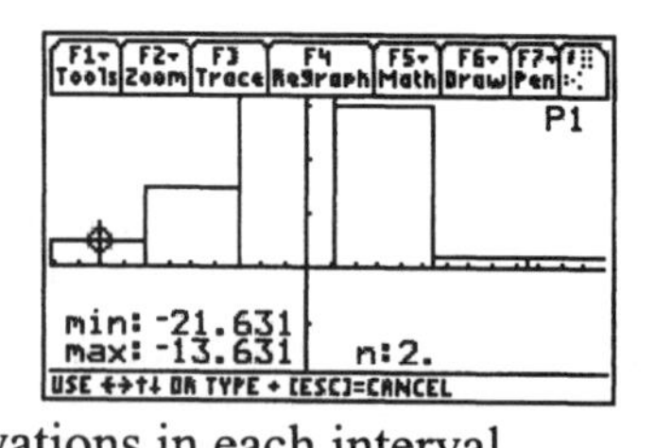

The easiest way to start displaying a histogram (or any statistics plot) is to press [F5] (Zoom Data). The resulting graph is seen at right. Notice the Y-axis penetrating one of the bars. The x-axis "floats" a little way up from the bottom of the screen. This is so that values as seen in the next picture do not interfere with the plot. This picture shows one bad point of the TI-89. It does not always get the windowing correct. We don't see the full heights of the bars. We'll change that later.

You will want to see just what the graph shows. To do this, press [F3](Trace). A blinking cursor will show in the first bar at the left of the graph. At the bottom of the screen the minimum value included in the bar, maximum value for the bar, and number of observations in the bar will be displayed. This bar goes from -$21.631 to -$13.631. There are two observations in this interval. Pressing the right arrow key ([▶]) will allow you to continue through the graph seeing the interval ranges and numbers of observations in each interval.

At this point, we can see the distribution of Enron stock prices appears to be unimodal and relatively symmetric (bars fall roughly equally from the center peak.) We see the center is around a price change of $0 with 38 observations in that bar; the worst change was a loss of $17.75 and the biggest gain was less than about $26.37. We'd really like to see the whole graph, however.

There is another downfall to using simply Zoom Data for histograms. Look at the intervals. They really do not make sense in a natural way. We'd like to fix this.

Manipulating Windows

To force particular minimums, maximums and scaling we will press [◆][F2] (Window). This displays the screen at right. Notice the xmin was the smallest value shown on the plot; xmax is the largest. ymin and ymax are analogous. xscl and yscl are the distances between axis "tick marks".What we really want to change are ymin and ymax so the entire graph can be seen.

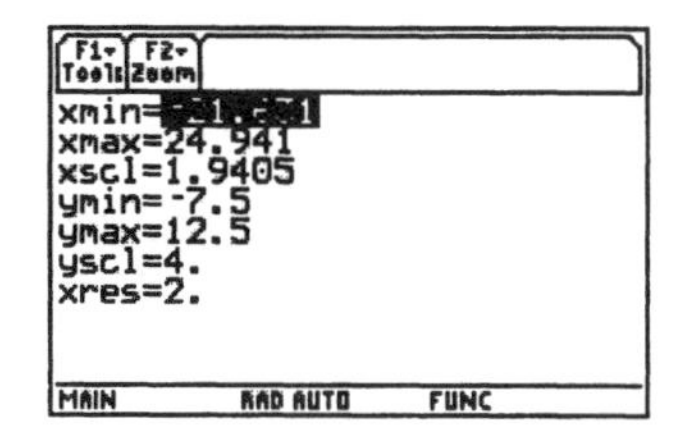

Change xmin to –18 (just slightly smaller than the smallest data value) ymin to –12 and ymax to 48. Why must ymax be so large? The "tabs" at the top of the screen will hide some of the plot unless it is sized on the roomy side. Ymin is set low so that the legends which appear after pressing Trace don't obscure portions of the graph.

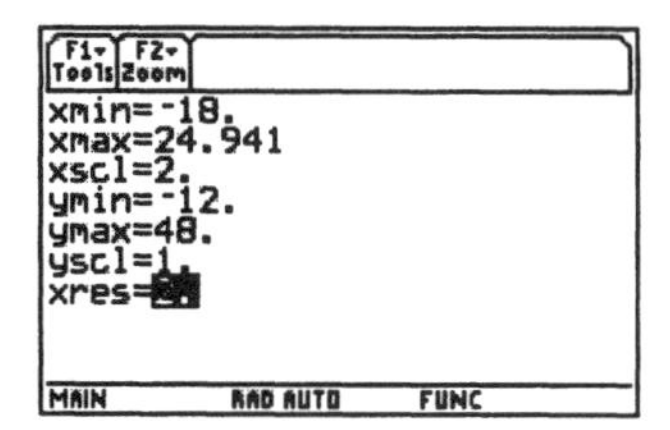

To display the new graph, press [◆][F3] (Graph). This looks better, but the bars seem rather wide. Let's make the bar width narrower.

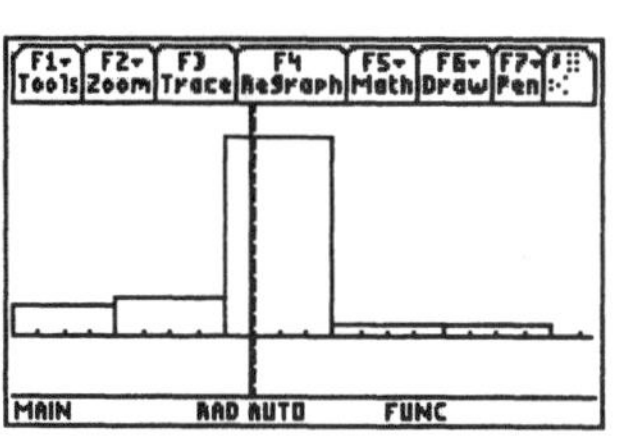

Press [◆][APPS] to return to the Statistics Editor, then go back to the plot definition menu. Change the bucket width to 5, then press [◆][F3]. The new graph is at right. We still see the strong central peak, but now it looks like the right side of the graph is somewhat longer than the left. We can play with other settings, just to see how the shape might change.

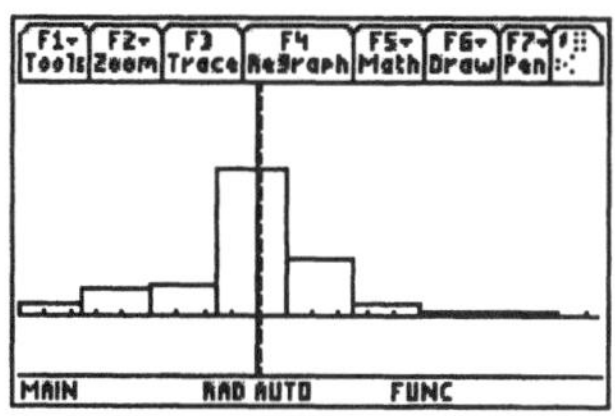

In this graph, the bucket width was set to 3. ymax is 24.The distribution still looks unimodal, but somewhat right skewed. Notice the gap (two bars worth) on the high end. It looks like we have an outlier.

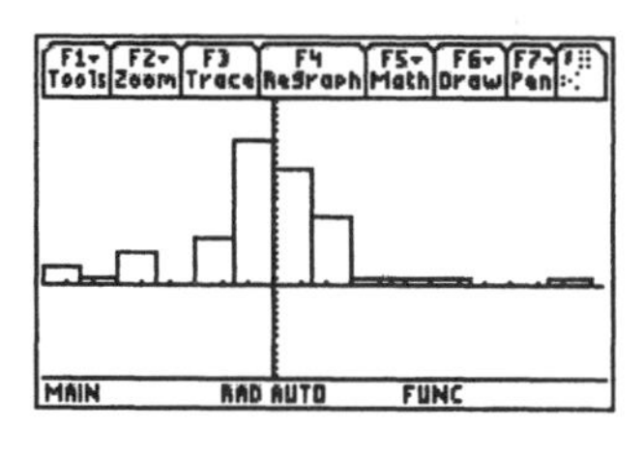

It's possible to have too many bars. Here, the bucket width has been set to 1; ymin is –8 and ymax is 12. (These were changed for picture resolution.) We're starting to lose the forest for the trees.

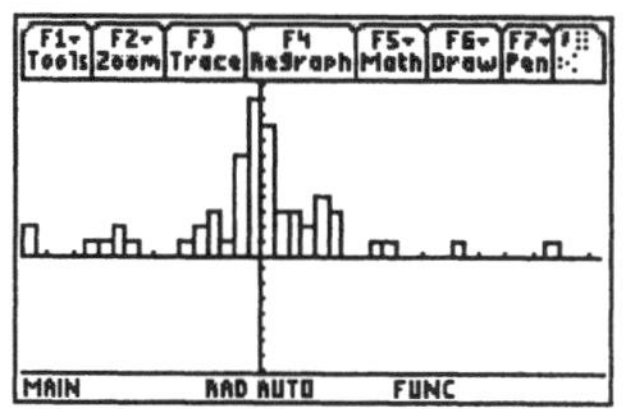

How many bars and where are somewhat personal judgement. Your instructor may give you some guidelines. One "rule of thumb" for many years was to have somewhere between 5 and 20 bars; for most data sets dividing the number of observations by 5 gives a good estimate of how many bars will give a decent picture.

"Printing" the Picture

Unfortunately, calculators do not have printers. To make a hard copy of the graph once you are satisfied, use the [TRACE] key to examine the entire graph. Make a picture of the histogram, clearly labeling each axis and giving the graph a title. Remember that the intervals given are the endpoints of the intervals. Label them as such. When you are finished, you should have a picture like the one at right.

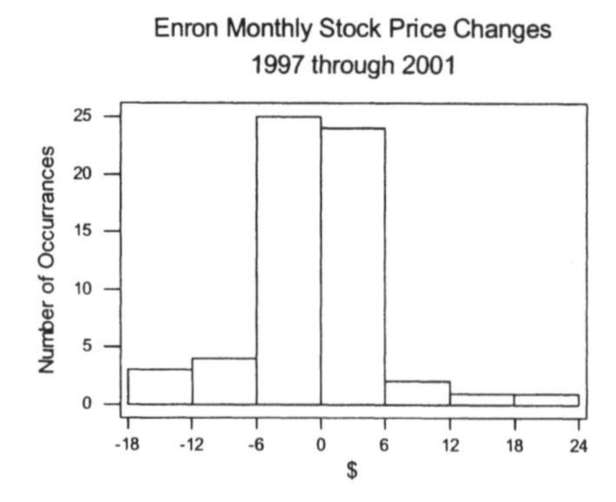

Histograms with Frequencies Specified

Sometimes data are given in the form of tables with both the data value and the number of times each value was observed. The frequency table below shows the heights (in inches) of 130 members of a choir. Entering 130 numbers could be tiresome, but there is a way to use the counts given.

We want to make a histogram to display this distribution. Enter the heights in one list and the observed counts in a second list. (We will put the heights in list1 and the counts in list2.) Make sure the lists are the same length, and that data values match with the given counts.

Height	Count
60	2
61	6
62	9
63	7
64	5
65	20
66	18
67	7
68	12
69	5
70	11
71	8
72	9
73	4
74	2
75	4
76	1

The first part of the lists looks like this.

list1	list2	list3	list4
60	2	------	------
61	6		
62	9		
63	7		
64	5		
65	20		

list3[1]=
MAIN RAD AUTO FUNC 3/6

From [F2] (Plots) we will define a new Plot 1 as at right. Notice x list is list1 (where the actual values are) and Freq is list2 (where the counts are) we have also toggled the Use Freq and Categories? to YES. We have also set the bucket width to 2 (again, this is judgement).

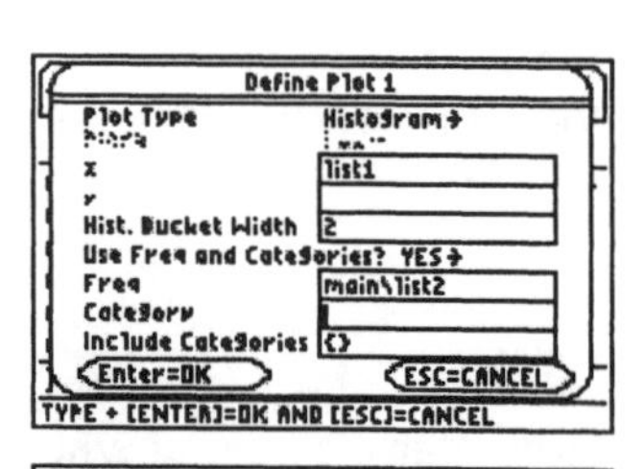

Pressing [F5] does not show the entire graph again. Changing the window settings (press [♦][F2] to get to the window screen) to use xmin = 60, xmax = 78, xscl = 2, ymin = -8, ymax = 28, yscl = 1 and then pressing [♦][F3] (Graph) gives the histogram at right. We see the distribution is unimodal and right-skewed. The visual center is around 68 inches.

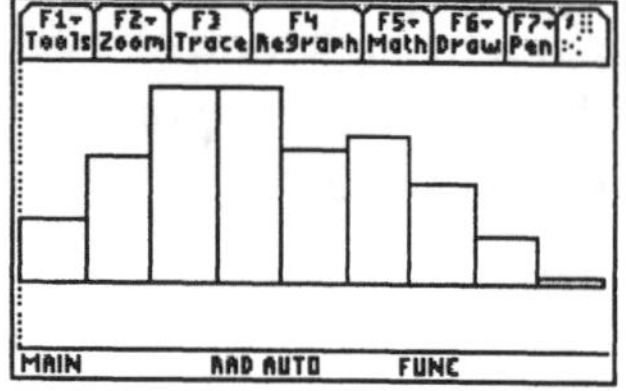

Histograms to Compare Distributions

We'd like to compare the distributions of two historically great baseball hitters: Babe Ruth and Mark McGwire. We have the following information on the numbers of home runs hit by Babe Ruth for 1920 through 1934 and for McGwire from 1986 to 2001.

Ruth: 52 59 35 41 46 25 47 60 54 46 49 46 41 34 22

McGwire: 3 49 32 33 39 22 42 9 9 39 52 34 24 70 65 32 29

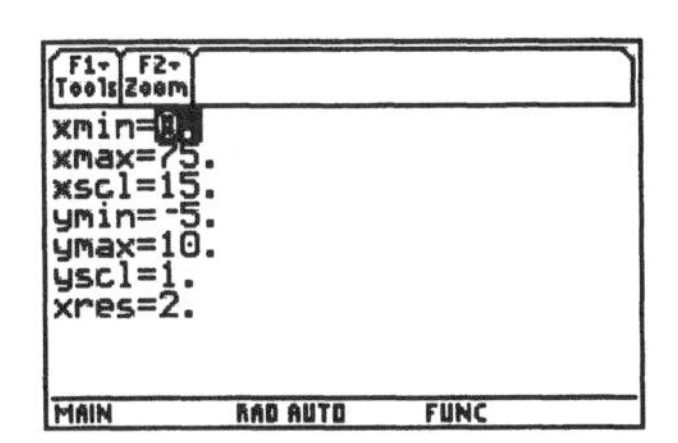

In order to adequately compare the distributions visually, they need to be on the same scaling (otherwise you're comparing apples and oranges.) Press ◆F2 (Window). Both the smallest and largest values occur in McGwire's distribution: 3 and 70. We also need a reasonable number of bars. It seems reasonable to set `xmin` to 0, `xmax` to 75. The bucket width on the plot definition screen has been set to 15. The window settings used are at right. The properly scaled plots are below.

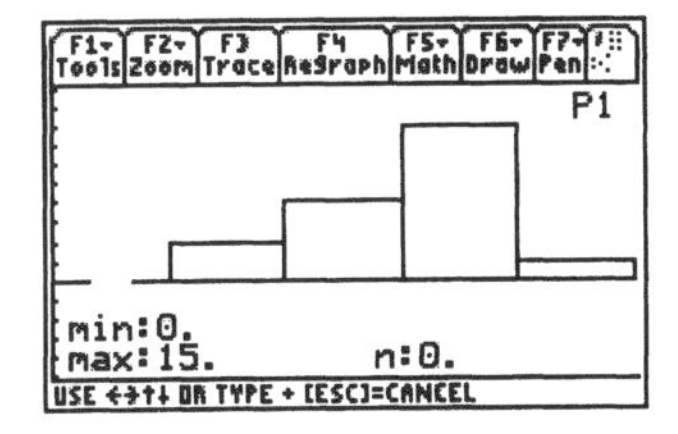

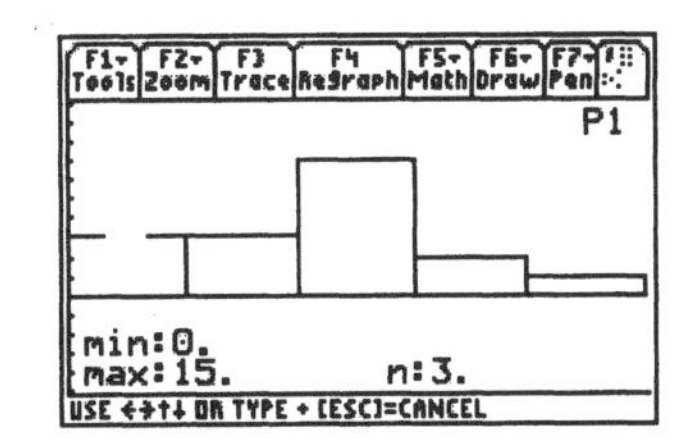

From these graphs it is easy to see that Ruth was the more prolific hitter. While McGwire had four years in which he hit more than 45 home runs, Ruth had 9. McGwire also had three seasons with fewer than 15 homers (due either to injury or his first, incomplete season).

Time Plots

Many variables are often measured at different points in time (as was the Enron data). It's not enough just to picture the distribution in this case. Time is an important factor, and we will want to know what (if any) part it plays. To answer this question, we will do a time (series) plot of the data. By convention, these are connected scatter plots with time represented on the x-axis and the actual variable values on the y-axis. They are connected because this makes any pattern easier to see than if the data points were just shown by themselves.

In our case, the time variable really consists of months and years, but TI calculators cannot handle this type of data. It will suffice for our purposes to simply use an index of months, from 1 to 60 to represent the five years' data. The data are in `list1` and the time index has been entered into `list2`.

list1	list2	list3	list4
-1.44	1	------	------
-.75	2		
-.69	3		
-.88	4		
.12	5		
.75	6		

list3[1]=

MAIN RAD AUTO FUNC 3/6

Connected scatter plots are the second plot type on the plot definition screen (called xyline). The x list is the time indices and y list is the Enron values. There are several choices for marking the actual data points. For this type of data, we actually recommend the dot mark since others may obscure the plot. Once the plot has been defined, it can be displayed by pressing [F5] (no need to fix windows here).

Here is the Enron Time Plot. We can clearly see that although there were fluctuations in price throughout the five-year period, these became much more dramatic (volatile) toward the end. In addition, it is clear that the price suffered steep declines toward the end.

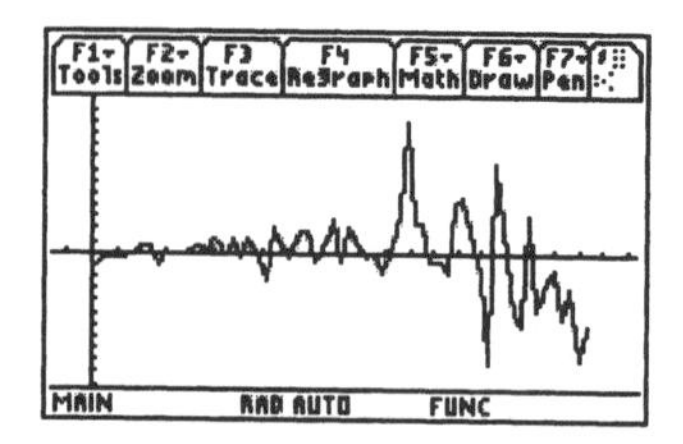

What can go wrong?

Help! I can't see the picture!

Seeing something like this (or a blank screen) is an indication of a windowing problem. This is usually caused by pressing [GRAPH] using an old setting. Try pressing [F2][9] (Zoomdata) to display the graph with the current data. Alternatively, go to the Window screen [2nd][F2] to adjust the window settings.

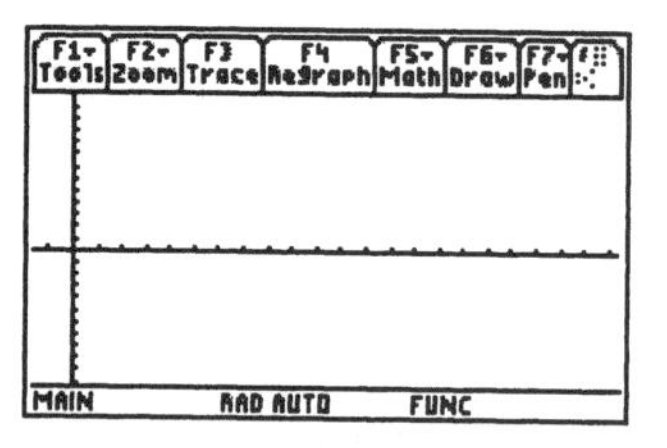

What's that weird line (or curve)?

There was a function entered on the [Y=] screen. The calculator graphs everything it possibly can at once. To eliminate the line, press [◆][F1] (Y=). For each function on the screen, press [CLEAR] to erase it. Then redraw the desired graph by pressing [◆][F3](Graph). Alternatively, go to the plots menu ([F2] from the Statistics Editor) and select choice 4: FnOff.

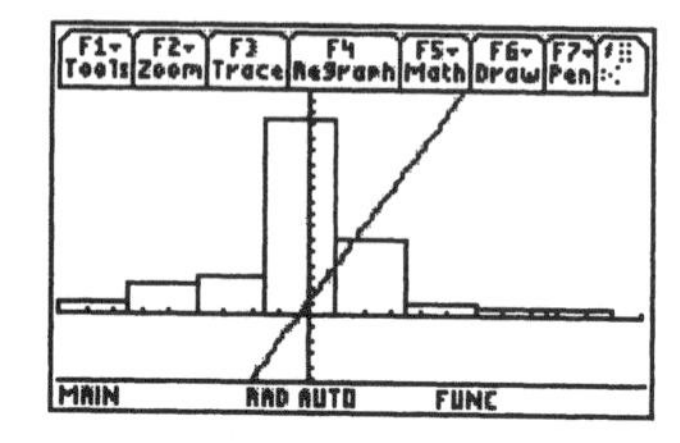

What's a Dimension Mismatch?

This common error results from having two lists of unequal length. Here, it pertains either to a histogram with frequencies specified or a time plot. Press [ESC] to clear the message, then return to the statistics editor and fix the problem.

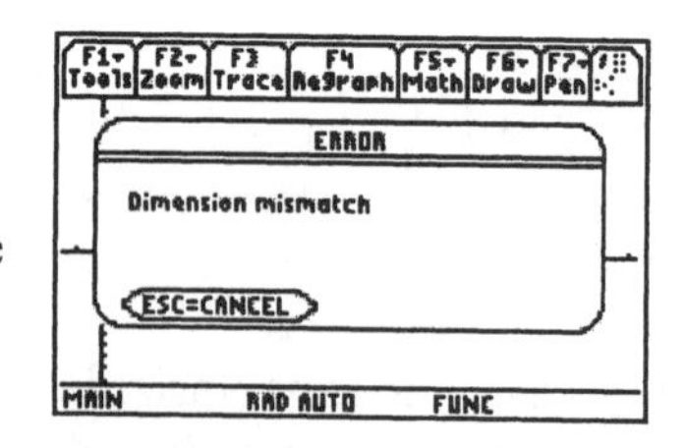

What's Dimension?

This problem is generally caused by reference to an empty list. Check the statistics editor for the lists you intended to use, then go back to the plot definition screen and correct them.

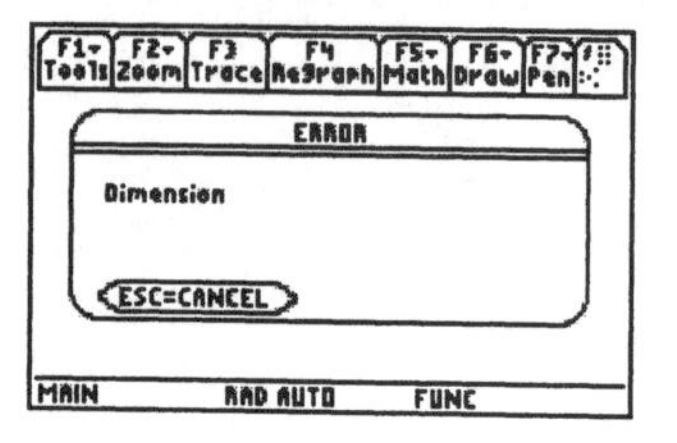

Plot setup?

This error is caused by having two stat plots turned on at the same time. What happened is the calculator tried to graph both, but the scalings are incompatible. Go to the Stat plots menu and turn off any undesired plot by moving the cursor to that plot, and pressing [F4].

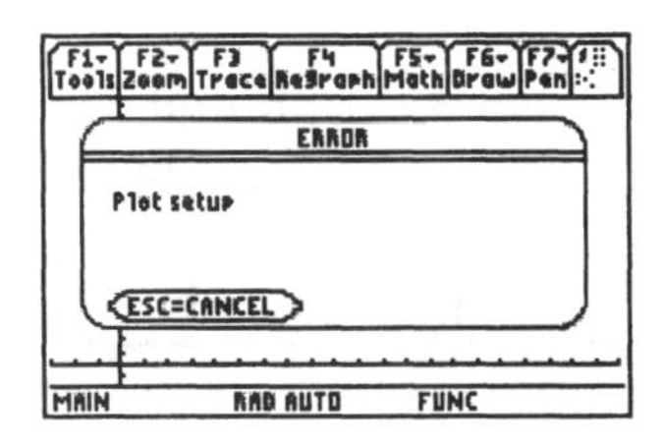

Chapter 16 – Describing Distributions Numerically

Remember the Enron data from the last chapter – monthly stock price changes for the five years preceding the crash. They are reproduced here for convenience.

	Jan	Feb	Mar	Apr	May	Jun	Jul	Aug	Sep	Oct	Nov	Dec
1997	-$1.44	-0.75	-0.69	-0.88	0.12	0.75	0.81	-1.75	0.69	-0.22	-0.16	0.34
1998	0.78	0.62	2.44	-0.28	2.22	-0.50	2.06	-0.88	-4.50	4.12	1.16	-0.50
1999	3.28	3.34	-1.22	0.47	5.62	-1.59	4.31	1.47	-0.72	-0.38	-3.25	0.03
2000	5.72	21.06	4.50	4.56	-1.25	-1.19	-3.12	8.00	9.31	1.12	-3.19	-17.75
2001	14.38	-1.08	-10.11	-12.11	5.84	-9.37	-4.74	-2.69	-10.61	-5.85	-17.16	-11.59

What is a typical "middle" value? What is the spread of the data? These are the questions addressed in this chapter. We will also meet a new statistics plot based on numerical summaries.

Calculating Numerical Summaries

To calculate the numerical summary statistics for a single variable, first enter them into a list. Here the data have been entered into `list1`. The first few values are shown at right.

Press [F4] (`Calc`). The menu at right will be displayed. The menu is organized so that the most often used options are at the top. Notice that `1:1-Var Stats` is highlighted. Press [ENTER] to select that option (or press [1]).

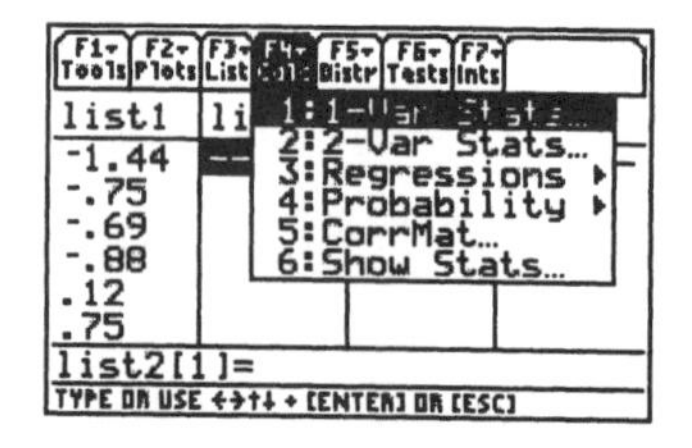

This is the input screen. You need to tell the calculator which list is to be used as input. Press [2nd][-] (`VAR-LINK`). Move the cursor to the correct list and press [ENTER] to select it. Since each value in our list occurred once, leave `Freq` at 1. Press [ENTER] to execute the command.

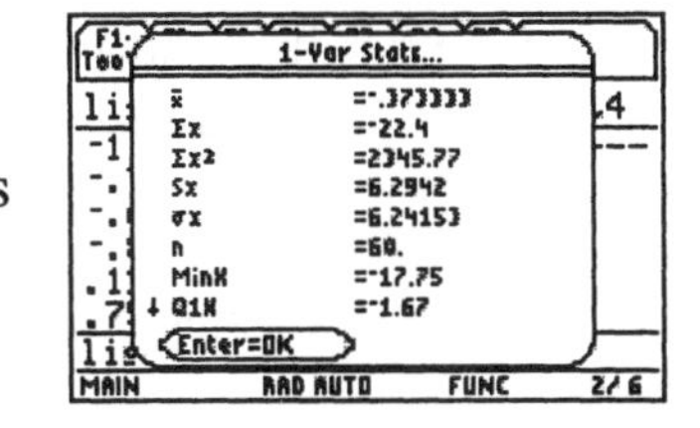

The first page of results is displayed at right. The arrow at the bottom left indicates more results are available and can be found by using ⊙. We first see the mean monthly stock price change is -\$0.3733333. The calculator does not know if the data you are using represent a sample or a population. It has only one symbol for the mean ($\bar{x}$). If your data represents a population, you should report the mean using proper notation, and call it μ. The two values displayed next are the sum of the data values and the sum of squared data values. These are intermediate quantities used in computations, and are generally not of interest. Two different standard deviations are also reported as measures of spread. $Sx = \sqrt{\dfrac{\sum (x_i - \bar{x})^2}{(n-1)}}$ is the sample standard deviation and σx is the population standard deviation (the formula is the same, except the divisor is n). This data does not represent all possible monthly stock changes for Enron, so we will use *Sx* (6.2942) as the standard deviation. Which is the correct value to use depends on data you have – is it from a sample or is it for a population? We also see the number of items in the data list, n, is 60. The last two entries begin the five-number summary.

One thing to bear in mind is that calculators (and computers) will use (and report) many more digits than really make sense to use. It comes from division (in which, as we know, things don't always come out evenly) and taking square roots (which generally aren't whole numbers). How many digits to report should be decided by your instructor, but a good rule of thumb is to report one more place than in the original data. Our data was in dollars and cents, so we'll use three decimal places. Also, since we don't have all the possible monthly stock price changes for Enron, we will report $\bar{x}$ = \$-0.373 and s = \$6.294.

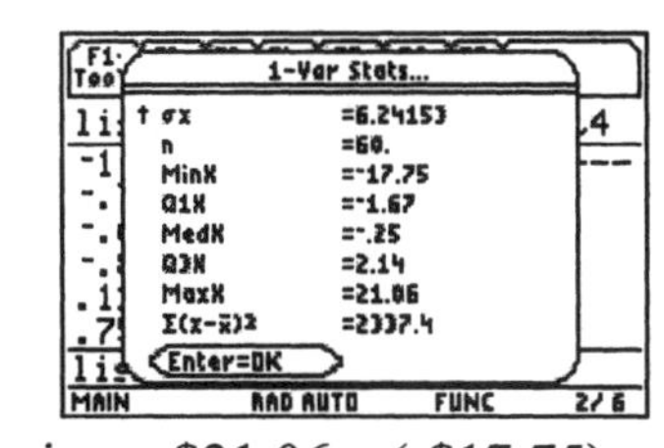

Using the down arrow, we find the five-number summary. The median (another measure of center) is -\$0.25, which is close to the mean in this data set (as it should be since the data were roughly symmetric). We can use the other values in this summary to compute two other measures of spread: the Interquartile Range (IQR) which is the spread of the middle half of the data, and the Range. The IQR is $Q_3 - Q_1$, or \$2.14 – (-\$1.67) which is \$3.81. This means the central half of the data had a spread of \$3.81. The Range is max – min, or \$21.06 – (-\$17.75) which is \$38.81.

Statistics for Tabulated data

In the last chapter we looked at the distribution of heights of members in a choir. They were presented in a table of heights along with how many choir members there were of a given height. With these data in lists `list2` and `list3`, we would like to know the average height for the choir.

Just as before, press [F4](`Calc`), press [ENTER] to select choice `1:1-Var Stats`. Now, change the entry in `Freq:` to refer to the list of counts (`list3`). Pressing [ENTER] gives the results desired.

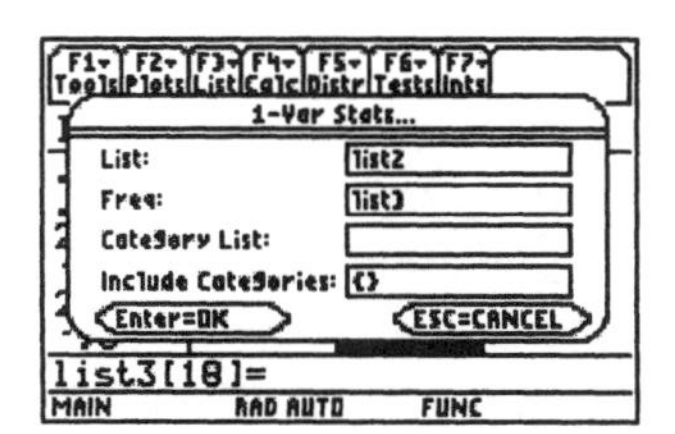

Here are the results. The average (mean) height for the choir members was 67.1 inches. The standard deviation (assuming these are not all the members possible for the choir) is 3.8 inches. Paging down, we find the median height was 66 inches. It is not surprising the median would be somewhat less than the mean for these data since the histogram indicated a right skewed distribution.

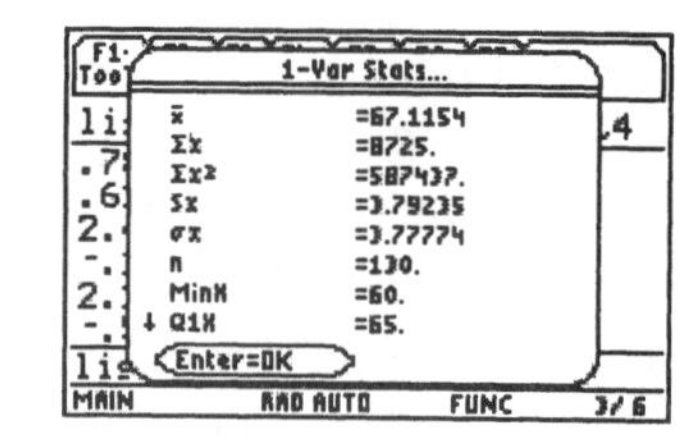

Boxplots

Box plots (sometimes called box-and-whisker plots) are another way of picturing a distribution. Unlike histograms, they are based on definite values and are not subjective. However, as no plot is perfect, they can hide some potential features such as bimodality. A good practice, since it is generally so easy, is to look at several plots. They all can show different things.

There are two types of boxplots – the original which is based on the five-number summary (min, Q_1, median, Q_3, and max) and a "modified" boxplot which has an objective criterion to identify outliers. Both types of plots divide the data into fourths – a "whisker" for the bottom and top quarters of the data, and a box for the middle half, with the median indicated inside the box. We always recommend using the modified boxplot, but your instructor may suggest otherwise.

As always, begin with data in a list. We will begin with the Enron data already examined. The data are in `list1`. Press [F2] (`Plots`) then [ENTER] to select `Plot Setup`, then [F1] to select defining `Plot1`.

Press the right arrow key to change the plot type. Notice there are two choices for boxplots – `3:Box Plot` which is the boxplot where outliers are not identified and `5:Mod Box Plot` which does identify outliers. We will look at both to see the difference between the two.

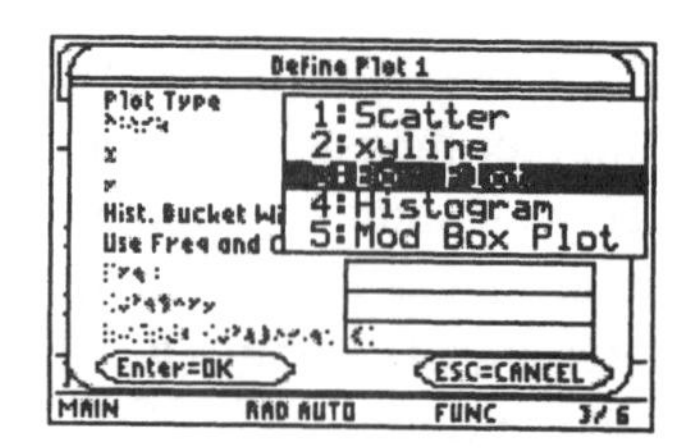

Move the cursor to highlight `3:BoxPlot`. Make sure `x` is changed to `list1` (press [2nd][-] to get to the `VAR-LINK` menu for list selection); also make sure `Use Freq and Categories` is set to No. Press [ENTER] to complete the plot definition.

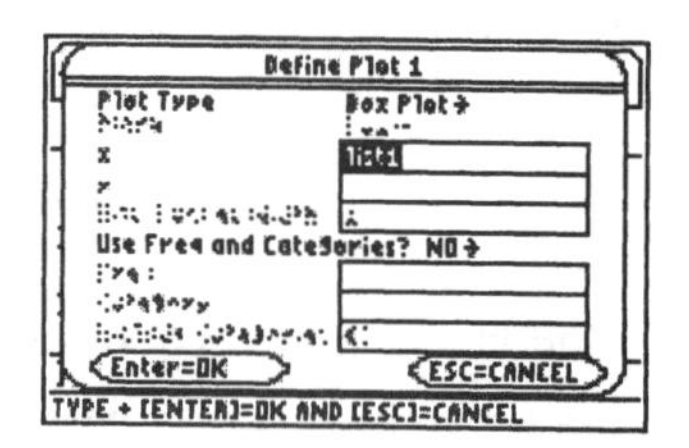

Press [F5](`ZoomData`) to display the graph. Here we see a short central box, indicating the middle half of the data is close to 0, and long relatively even whiskers. The median (-0.25) is very close to zero and is hard to distinguish. The indications from the plot are that the distribution is symmetric. Pressing [F3] (`Trace`) and using the right and left arrows will allow you to move around the graph, locating the median, quartiles, min, and max.

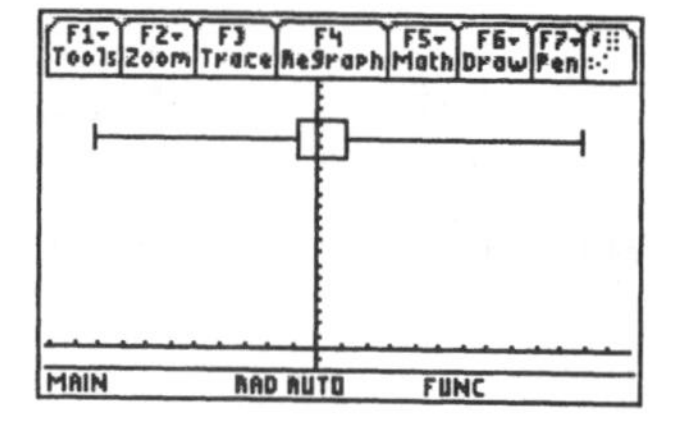

Return to the Plot definition screen and change the plot type to a modified boxplot identifying outliers (Choice 5). You may also choose your preference of Mark for any outliers which may be found. In this instance, we do not recommend the Dot choice (it's too hard to see the single pixels). Pressing [F5] after completing the definition will display the plot at right. The central box has not changed; the whiskers are shortened and many points on either end have been flagged as outliers by the $Q_3 + 1.5*IQR$ and $Q_1 - 1.5*IQR$ criteria.

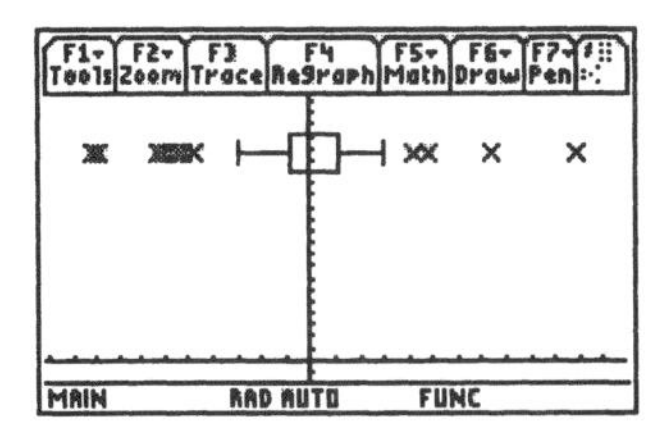

Boxplots with Tabulated Data

Reconsider the data on heights of members of a choir. According to the histogram, this was somewhat right-skewed. What will its boxplot look like? With heights in list2 and frequencies in list3, the plot definition screen looks like that at right.

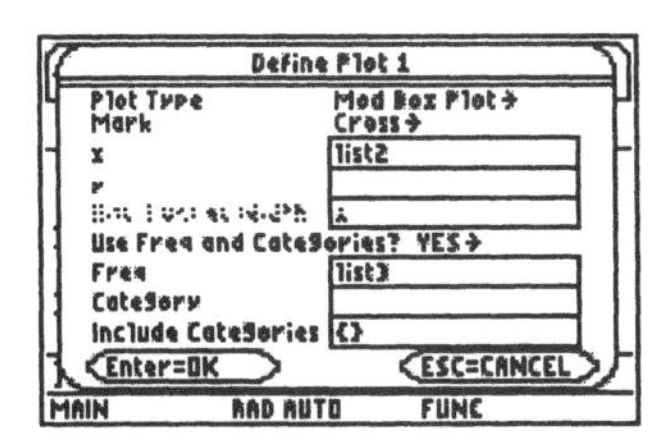

After completing the definition and pressing [F5] the graph at right will be displayed. Notice the median is not in the middle of the box; the right half of the data is longer than the left half (the data is right skewed) even though the two whiskers are relatively equal in length. Also, since we defined the plot to identify any possible outliers, none are flagged, so these data have no outliers.

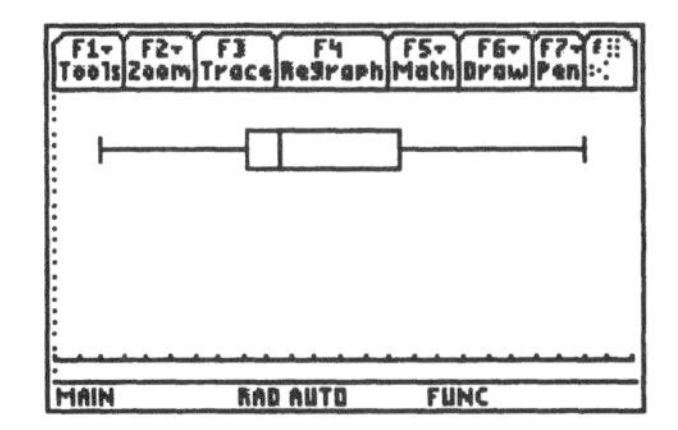

Boxplots to Compare Distributions

Boxplots are very useful in comparing distributions. This is one of the few exceptions to the rule about only one plot being turned on at a time. Several boxplots can be displayed at once. Displaying boxplots side-by-side is a good visual comparison of distributions.

Recall the data for Babe Ruth and Mark McGwire. How would these distributions look displayed together as boxplots? The data have been entered into list4 (Ruth) and list5 (McGwire). We define Plot 1 to use Ruth's home run values as at right.

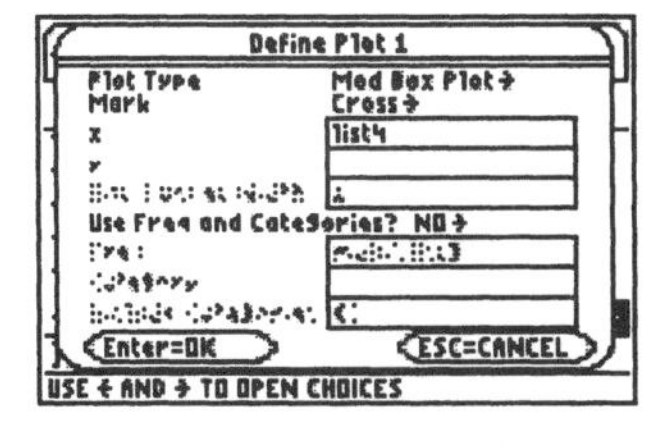

Returning to the StatPlots menu, we will arrow down to Plot 2, and define it to use McGwire's numbers as at right.

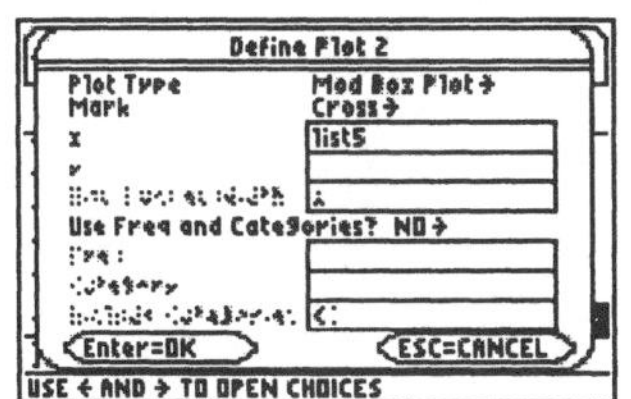

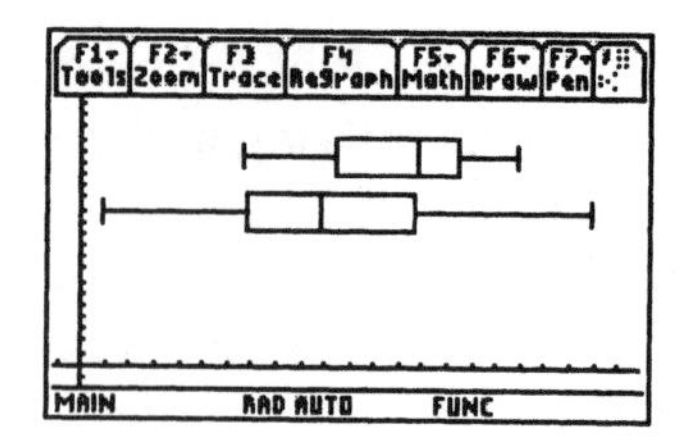

Pressing [F5] gives the display at right. The top plot is Ruth's home run distribution; the bottom is McGwire's. (Pressing [F3] will identify each plot; to move from one to the other, press the down and up arrows.) Neither distribution has outliers. Ruth's is much less variable than McGwire's and is skewed left. McGwire's distribution appears rather symmetric.

Chapter 17 – The Standard Deviation as a Ruler and the Normal Model

The standard deviation is the most common measure of variation; it plays a crucial role in how we look at data. *Z*-scores measure standard deviations above or below the mean and are useful as measures of relative standing. Normal models are very useful as many random variables (at least approximately) follow its unimodal, symmetric shape. TI calculators can replace the need for paper tables of the normal distribution.

Z-Scores

The *z*-score for an observation is $z = \dfrac{(obs - \bar{y})}{s}$, where *obs* is the value of interest. Positive values indicate the observation is above the mean; negatives mean the value is below the mean. Calculating them is easy as long as one keeps in mind that the subtraction in the numerator must be done before the division. Calculators follow the arithmetic hierarchy of operations.

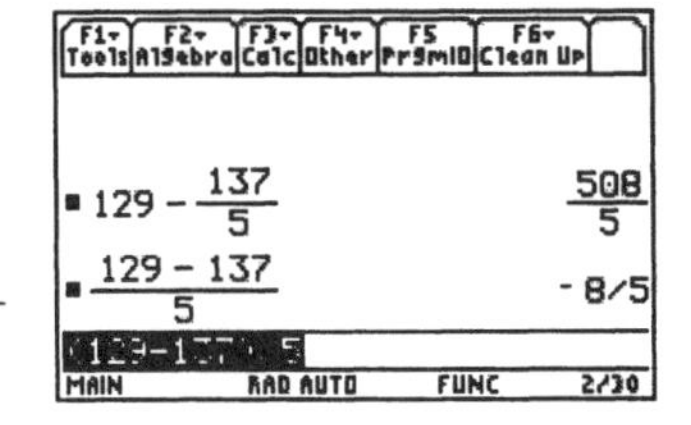

For example, in winning an 800m race during the Olympic women's heptathlon, Gertrud Bacher of Italy had a time of 129 seconds; the mean time for all runners in the race was 137 seconds with standard deviation 5 seconds. What is Bacher's *z*-score? Two examples of the calculation are at right. The first (incorrect!) indicates she was 101.6 standard deviations *above* the mean – unreasonable since she won the race, so had the fastest time. The problem is failing to perform the subtraction first or enclosing the numerator in parentheses. The second (correct!) calculation indicates Bacher's time was 1.6 standard deviations *below* the mean.

Working with Normal Curves

What proportion of SAT scores are between 450 and 600? SAT scores for each test (verbal and math) are approximately normal with mean 500 and standard deviation 100, or N(500, 100). There are two ways to answer this question with a TI-89 calculator. One will draw the curve; the other just answers the question. Both start at the same place: the Distributions menu.

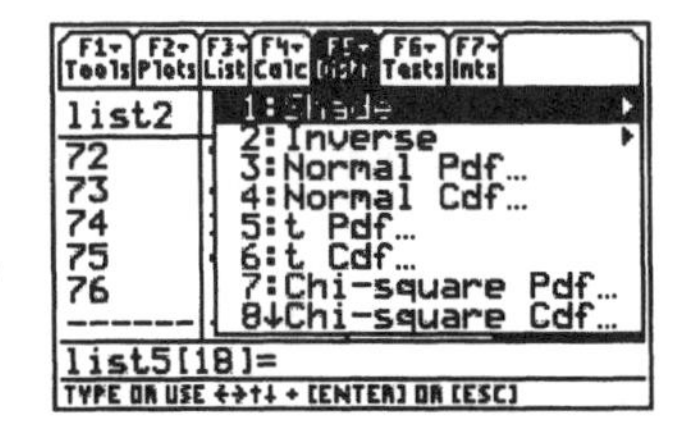

From the Statistics Editor Screen, Press [F5] (DISTR) The screen at right will appear. Notice the arrow pointing down at the bottom left. There are more distributions which can be used; more will be said about some of them later. The first two choices are Shade and Inverse. Notice these also have sub-menus(indicated by the ▶). Shade will draw pictures of the distribution curves and highlight the relevant portion; Inverse is used for finding quantiles of a distribution.

First, let's just answer the question. The menu option to select is `4:Normal Cdf`. Either press the down arrow and then [ENTER] or just press [4]. You will be asked to supply four parameters: the low end of the area of interest, the high end, the mean (which is 0 for the standard normal distribution) and finally the standard deviation (1 for standard normal). Press the down arrow key to move between boxes.

A score of 450 is 0.5 standard deviations below the mean, so its *z*-score is –0.5. A score of 600 is one standard deviation above the mean; its *z*-score is 1. These have been entered as the low and high ends of interest. Since we're working with standardized values, the mean is 0 and the standard deviation is 1.

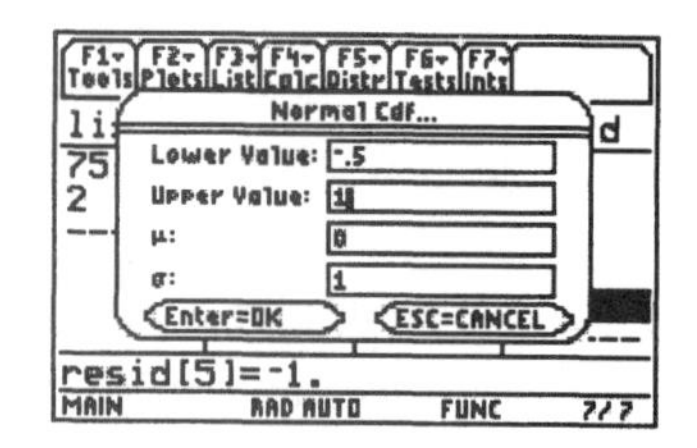

Here are the results. We see that about 53.3% of all scores on the SAT will be between 450 and 600.

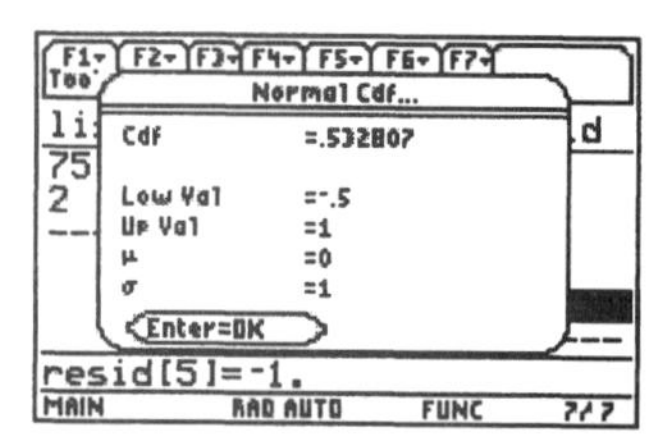

To find the area and have it shaded we will do the following. Press [F5] then since option `1:Shade` is highlighted, press the right arrow key. This gives a choice of 4 options. Press [ENTER] to select Shade Normal.

Specify the lower and upper values, then the mean and standard deviation. Select YES in answer to `Auto-scale` as in the screen at right.

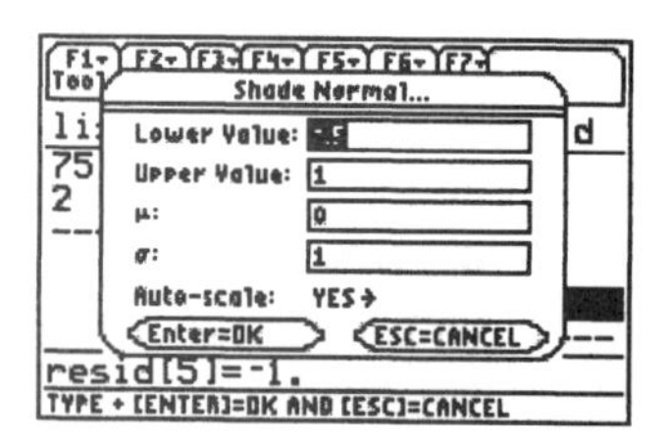

Pressing [ENTER] will display the properly shaded Normal Distribution, along with the probability.

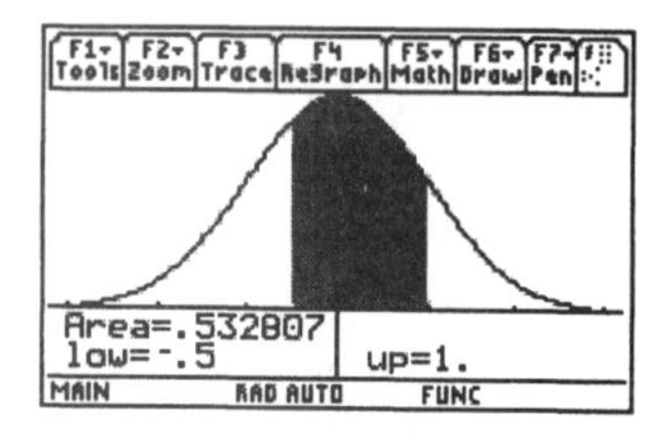

Let's look at another example. A cereal manufacturer makes boxes labeled as 16 ounces; but the boxes are actually filled according to a normal model with mean 16.3 ounces and standard deviation 0.2 ounces. We want to know what fraction of all boxes will be "underweight," that is, contain less than the advertised 16 ounces.

Strictly speaking, Normal models extend from $-\infty$ to ∞ (negative infinity to infinity). On the calculator, ∞ is represented as 1e99 (10^{99}). To enter this, one presses [1][EE][9][9], but practically, any "very large" negative number (say, -99) will work for $-\infty$ and any large positive number (say 99) for ∞ since there is so little area more than 99 standard deviations above the mean.

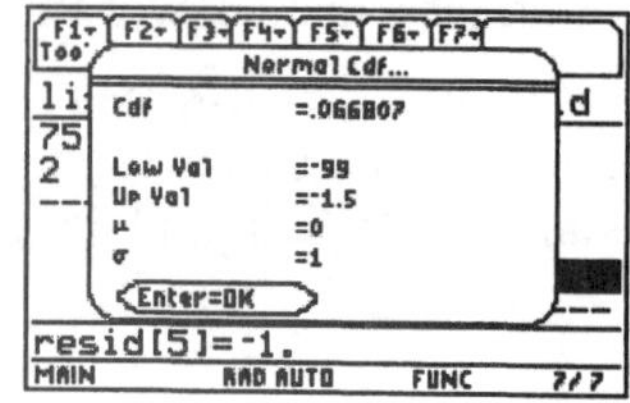

We want to know what fraction of all boxes are less than 16 ounces, so the low end of interest is $-\infty$; the upper end of interest is the z-score for 16 which is –1.5. The results are at right. We see that about 6.7% of all boxes of this cereal should be underweight.

Working with Normal Percentiles

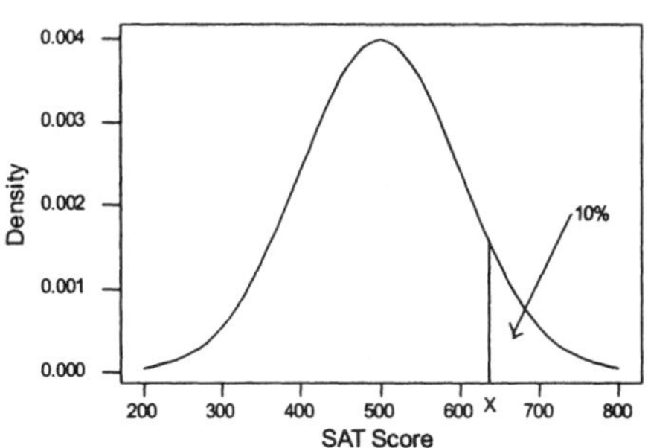

Sometimes the area under the curve is given and the corresponding value of the variable is of interest. For example, in the SAT model used before, how high must a student score to place in the top 10%? In a sketch of the normal curve, the unknown value, we'll call it *X*, separates the top 10% from the lower 90%.

This is the inverse situation from that we've just explored. On the DISTR menu, press the down arrow to 2: Inverse then press the right arrow and select 1: Inverse Normal. We will first find the *z*-score corresponding to the SAT score which separates the top 10% from the bottom 90%. The parameters for this command are area to the left of the point of interest (.90 or 90%), the mean (0) and standard deviation (1). Press ENTER to execute the command. The *z*-score of interest is 1.28. To be in the top 10%, your score must be 1.28 standard deviations above the mean. $z = (x - \mu)/\sigma = 1.28$. Filling in the values for the mean and standard deviation, we have $(x - 500)/100 = 1.28$. After doing the algebra, we see that a score of 628 will put a person in the top 10% of all SAT scores; practically since scores are reported rounded to multiples of 10, a score of 630 is needed.

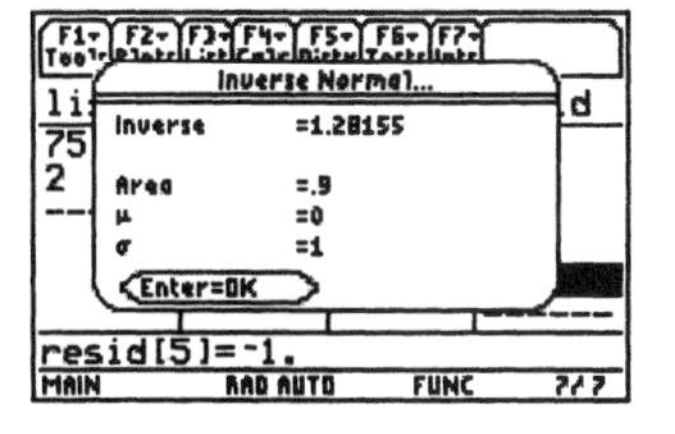

Here is another example. The cereal company's lawyers are not happy with 6.7% of boxes being underweight. They want at most 4% to be underweight. What mean must the company reset its machines to in order to achieve this target? We'll use Inverse Normal for a standard normal model to find the z-score corresponding to 4% of the area below this value, then use some algebra to solve for the unknown mean.

The *z*-score of interest is –1.75.

$z = (x - \mu)/\sigma = -1.75$

$(16 - \mu)/0.2 = -1.75$

Now multiplying both sides by 0.2, and subtracting 16 from both sides gives $-\mu = -16.35$ or $\mu = 16.35$. In order to achieve the target of no more than 4% of boxes being underweight, the machine will have to be set for an average of 16.35 ounces per box.

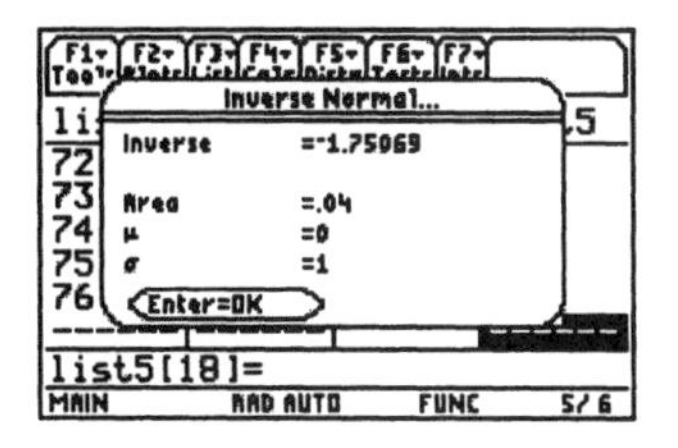

Is my data Normal?

It is one thing to assume data follows a normal model. When one actually has data this should be checked. One method is to look at a histogram – is it unimodal, symmetric and bell-shaped? Another is to ask whether the data (roughly) follow the 68-95-99.7 rule. Both of these might work well with a fairly large data set; however, there is a specialized tool called a normal probability plot that will work with any size data set. This plots the data on one axis against the *z*-score one would expect if the data were exactly normal on the other. If the data is normal this plot will look like a diagonal straight line.

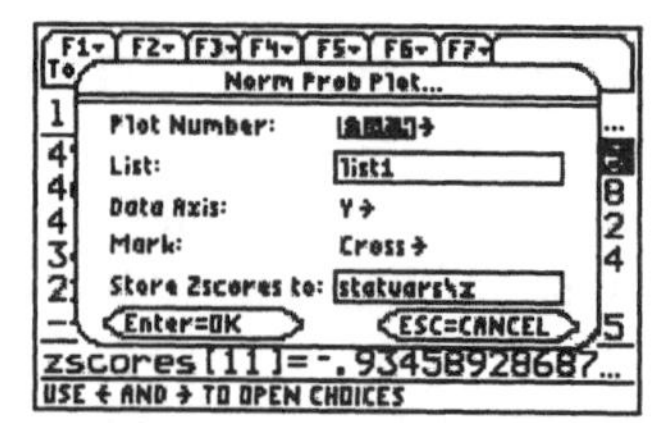

Recall the data on Enron stock price changes. They are in list1. Press [F2] (Plots) and select choice 2: Norm Prob Plot. The plot number defaults to one more than the last plot defined. Enter the list name to use (press [2nd][-] to get to the VAR-LINK screen). Notice you have a choice of having the data on either the *X* or *Y* axis. It doesn't really matter which you choose; many statistical packages put the data on the *x*-axis; many texts (including DeVeaux and Velleman) put the data on the *y*-axis. As we have seen before, you have a choice of marks for each data point. Select the one you prefer. The calculator will store the *z*-scores in a list. Just take the default here. Pressing [ENTER] calculates the *z*-scores. To display the plot, press [F2] again, check that all other plots are "turned off" (uncheck them by pressing moving the cursor and pressing [F4]) then [F5] to display the plot.

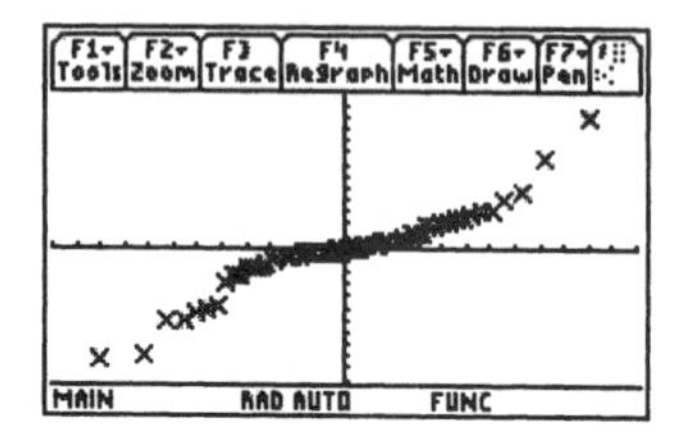

This graph is not a perfect straight line, and shows some stragglers (outliers) at each end just as we've seen before. I would hesitate to call this data normal.

Skewed distributions often show a curved shape. Data on the cost per minute of phone calls as advertised by Net2Phone in USA Today (July 9, 2001) to 22 countries were as follows:

7.9	17	3.9	9.9	15	9.9	7.9	7.9	7.9	7.9	8.9
21	6.9	11	9.9	9.9	7.9	3.9	22	9.9	7.9	16

We have entered the data in a list and have definee the normal probability plot as above.

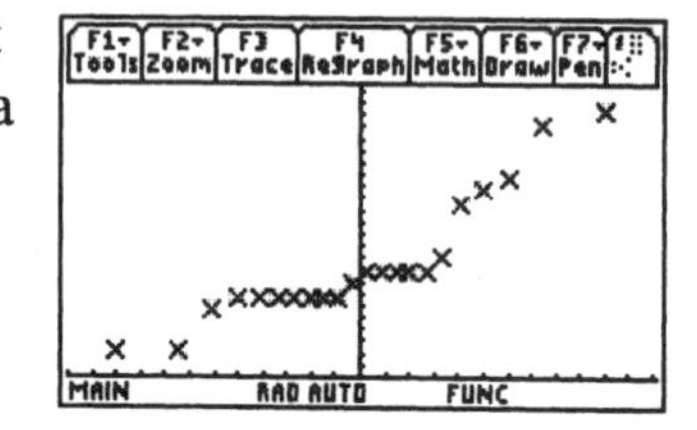

The plot obtained is at right. Not only does it show a general upward curve, it also displays something called *granularity*. This occurs when a particular data value occurs several times (as with 7.9 cents per minute which was in the list 7 times.)

What can go wrong?

How can the probability be more than 1?

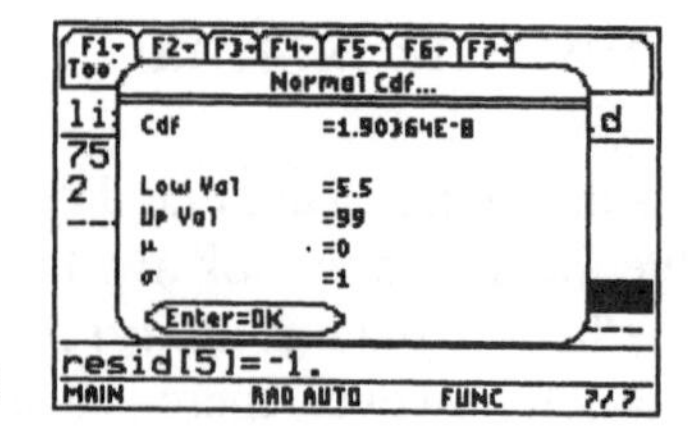

It can't. If the results look like the probability is more than one, check the right side of the result for an exponent. Here it is –8. That means the leading 1 is really in the eighth decimal place, so (rounding) the probability is 0.00000002. The chance a variable is more than 5.5 standard deviations above the mean (this would be a box of the cereal is more than 17.4 ounces) is about 0.000002%.

What's Err: Domain?

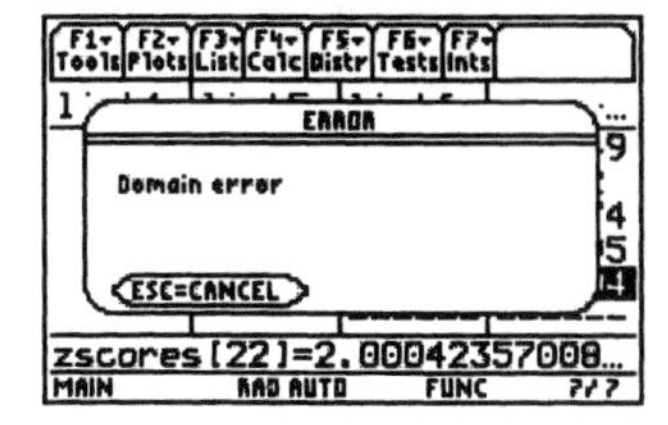

This message comes as a result of one of two problems: having entered the Inverse Normal command with an area greater than 1 or a Normal Cdf command with the low value greater than the upper value. (You wanted to find the value that puts you into the top 10% of SAT scores. For Inverse Normal problems, percentages must be entered as a decimal numbers. Pressing [ENTER] will return you to the input screen and allow you to correct the error.

Chapter 18 – Scatterplots, Correlation, and Regression

Are two numeric variables related? If so, how? Scatterplots and regression will answer these questions. Correlation describes the direction and strength of linear relationships. Linear regression further describes these relationships.

Here are advertised horsepower ratings and expected gas mileage for several 2001 vehicles.

Audi A4	170 hp	22 mpg	Buick LeSabre	205	20
Chevy Blazer	190	15	Chevy Prism	125	31
Ford Excursion	310	10	GMC Yukon	285	13
Honda Civic	127	29	Hyundai Elantra	140	25
Lexus 300	215	21	Lincoln LS	210	23
Mazda MPV	170	18	Olds Alero	140	23
Toyota Camry	194	21	VW Beetle	115	29

How is horsepower related to gas mileage? The first step in examining relationships is through a scatterplot.

Scatterplots

Here are the first few values for the horsepower ratings (in `list1`) and the gas mileage (in `list2`). It is important to enter these very carefully as they have been entered in the table, since the values represent a data pair for each vehicle type.

list1	list2	list3	list4
170	22		
190	15		
310	10		
127	29		
215	21		
170	18		

list3[1]=

MAIN RAD AUTO FUNC 3/7

Our supposition is that larger engines will get less gas mileage, so we will use the horsepower ratings as the predictor (X) variable and the gas mileage as the response (Y) variable.

To define the scatterplot, press [F2] (`Plots`) then select choice `1:Plot Setup`. Make sure unnecessary plots are cleared out by moving the cursor to highlight the plot and pressing [F3]. Press [F1] to start defining Plot 1. Press the right arrow to select the plot type. Press [ENTER] to select `1:Scatter`. Press the down arrow ([▼]) and select your choice of Mark for the data points. The author recommends against the last choice `5:Dot` as the single pixel tends to be too hard to see. Press the down arrow and select the list with the X values ([2nd][-] gets the `VAR-LINK` screen; move the cursor to the desired list and press [ENTER]). Press the down arrow and enter the list containing the response (Y). When finished, the plot definition screen should look like the one at right.

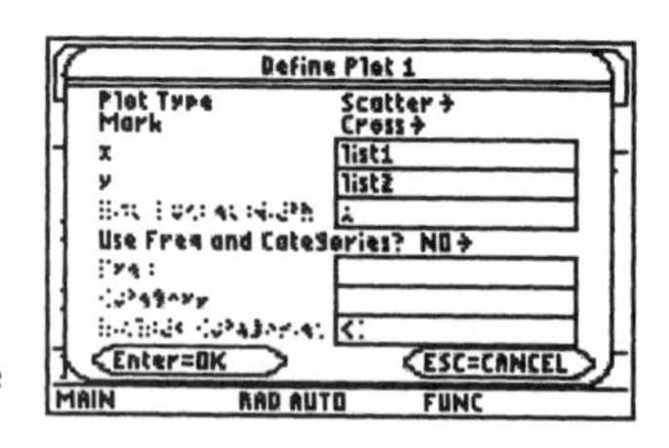

Press [F5] to view the plot. Here we see a generally decreasing pattern from left to right, supporting our initial idea. The pattern is generally linear; however, two points at the bottom right may be unusual; we'll examine those later.

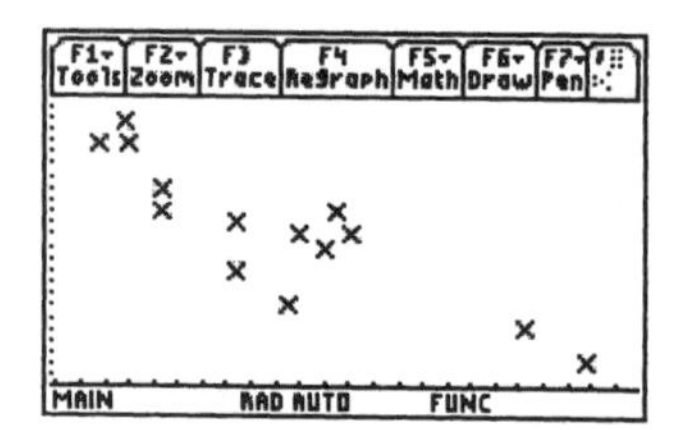

Regression and Correlation

We're now ready to examine the correlation between these two variables. However, the calculator will not give just the value of r; it's much easier computationally for it to do the whole thing at once and report all the values of interest.

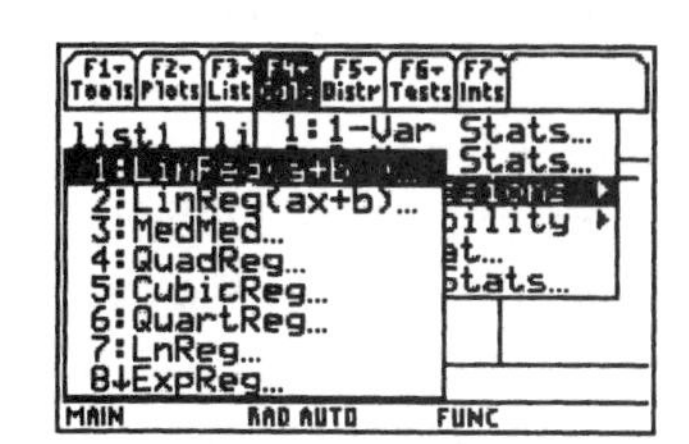

From the Statistics Editor, press [F4] (Calc). Arrow down to choice 3:Regressions and press the right arrow. Both choices 1 and 2 on this submenu are linear regressions The answers you get will be the same, but one must keep in mind the order in which the coefficients are used. Since statisticians usually prefer the constant term of the regression to come first (in case there are several predictor variables – multiple regression) we'll use choice 1. Press [ENTER] to select it.

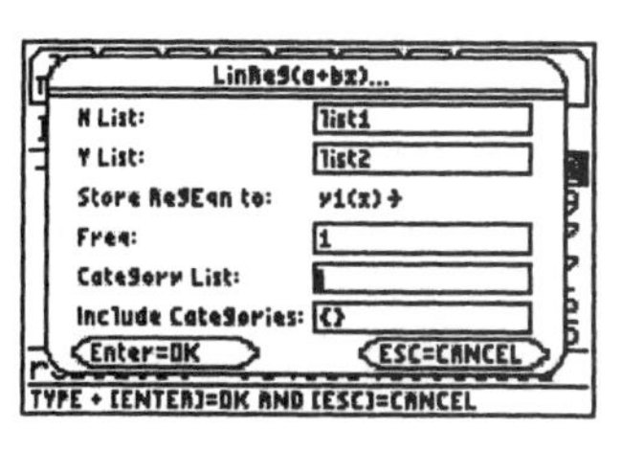

You will be presented with an input screen like those we have seen before. It asks for the list containing the X (predictor) variable; the Y (response) variable; and gives you the option of storing the equation of the line. With the right arrow here you can select none or a y-function. Since one usually wants to see the line plotted on the data graph, it is a good idea to select a function (usually y1(x)). Freq should be left at 1. My regression definition is at right. When finished with the definition, press [ENTER].

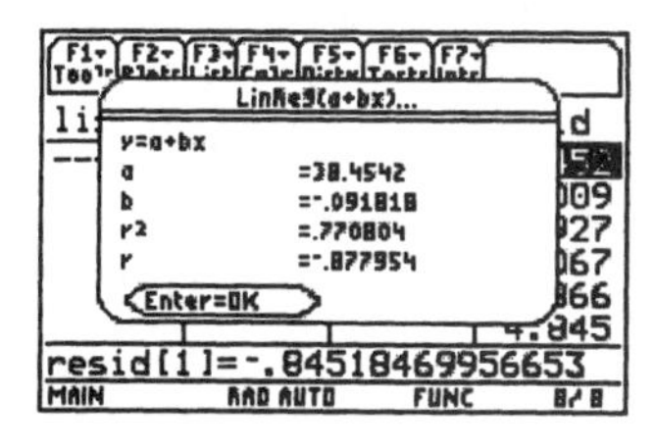

Once the command is executed, you should see the display at right. Notice the first line of the results displays the type of regression in terms of y and x. Regression lines should never be reported in these terms, but the calculator does not know what variables you are working with. This is really an aid to remind you where the coefficients a and b go in the equation.

Here, we have the following regression equation: $Mileage = 38.45 - 0.09 * horsepower$. Remember, this line represents the average value of gas mileage for a given horsepower rating, based on the model from our data. In addition, we see the correlation coefficient, $r = -0.878$ which indicates a strong, negative relationship. The coefficient of determination, $r^2 = 0.771$ (normally expressed as $r^2 = 77.1\%$) tells us that 77.1% of the observed variation in gas mileage (remember these values range from 10 to 31 mpg) is explained by the model.

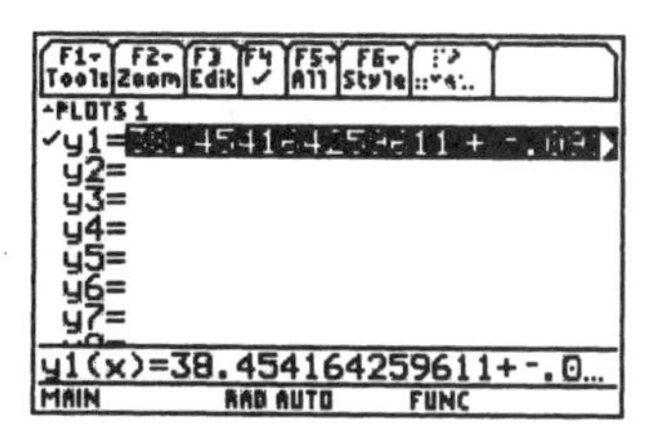

Pressing [◆][F1] (Y=) shows the first portion of the regression line as stored. It shows as checked, so it will be displayed on any graph.

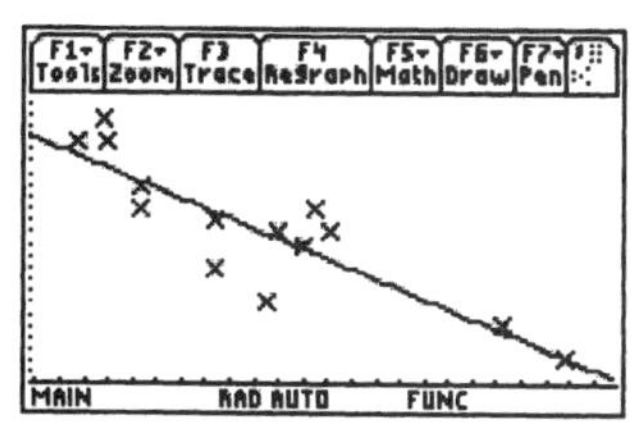

It would be nice to see how the line passes through the data; it should be roughly in the center of the data points. Press [◆][F3] (GRAPH), since there is no need to resize the window or redefine a plot. Sure enough, there's the line just as we expected. Notice that since no line will be perfect (unless $r = \pm 1$), some of the points are above the line, and some below. The distances between the points and the regression line are called residuals and their plots are used to examine the line for adequacy of the model.

Residuals Plots

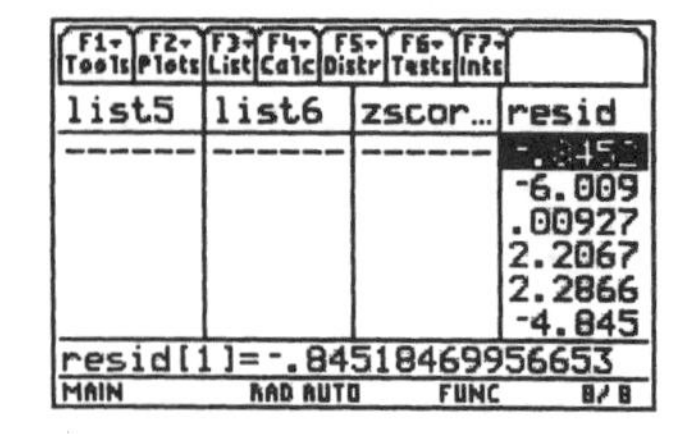

Residuals are defined to be the vertical distance from the data point to the regression line, in other words, $e_i = y_i - (a + bx_i)$ for each data point (x_i, y_i) in the data set. The e_i are the residuals. Returning to the Statistics Editor, we find a new list called `resid` has been created by the regression.

There are two main types of residuals plots which should be done to examine the adequacy of the model for any regression. The first plots the residuals against X (the predictor variable); the second is a normal probability plot of the residuals. In the first plot, we hope to see random scatter in an even band around the x-axis ($Y = 0$ line). Any departures from this are cause for reexamination of the model. In particular, curves may appear which are "masked" by the original scaling of the data; subtraction of any linear trend will magnify any curve. Another common shape that indicates problems is a "fan" in which plot either narrows from left to right or, alternatively, thickens. Either of the fan shapes means there is a problem with an underlying assumption: namely that the variation around the line is constant for all x-values. If this is the case, a transformation of either Y or X is usually necessary. Unusual observations (outliers) may also be seen in these plots as very large positive or negative residuals.

A residuals plot against X (the predictor variable)

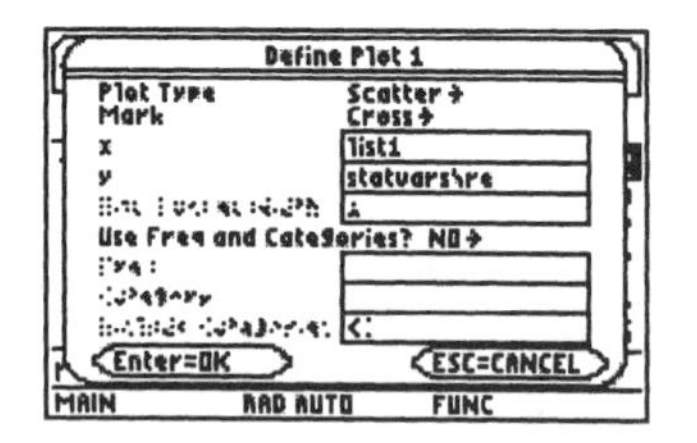

This is a scatter plot. From the plot definitions screen (press [F2][ENTER][F1] to define `plot 1`) define the plot using the original x-list of the regression (in out example `list1`) and the residuals list (in `VAR-LINK` arrow down until you find the `resid` list in the `Statvars` section) Press [ENTER] to complete the plot definition and return to the Plot Setup menu.

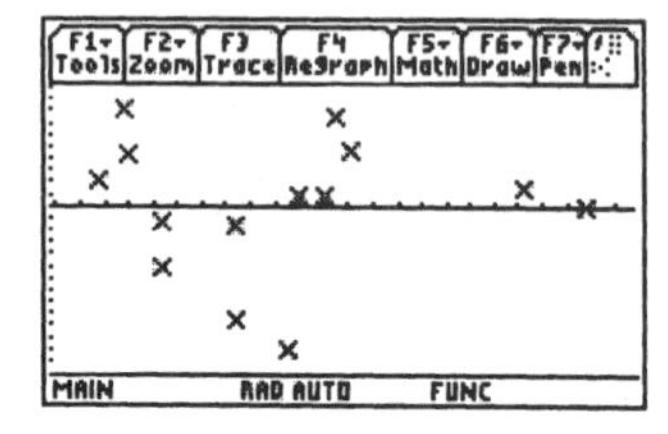

Pressing [F5] will display the residual plot at right. Looking at the plot, we see no overt curves, indicating a line appears to be an adequate model. This was a small data set; with these seeing non-constant variation can be difficult. There does not seem to be much of a problem except at the far right end of the graph, but there were few points there, so it's hard to tell. There do not seem to be any extremely large positive or negative residuals (outliers).

Normal Probability Plots of Residuals

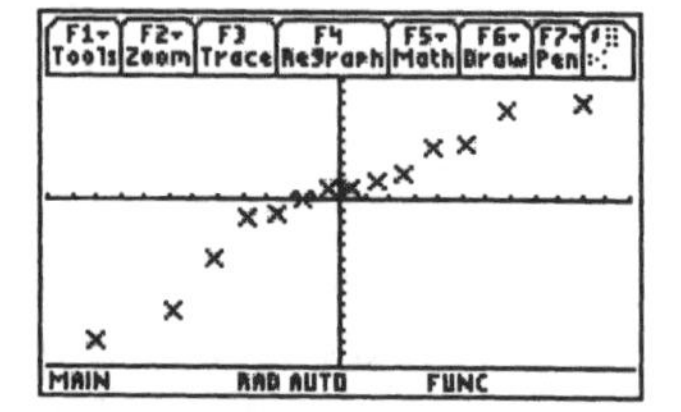

The second plot which should be done is a normal probability plot, since there is an underlying assumption the residuals have a normal distribution. This assumption will be used later in inference for regression. Normal probability plots were discussed in Chapter 17. Remember that the data list is the list of residuals. We're looking for a (approximately) straight line. Here, the pattern is (very) roughly linear, indicating no serious problems with this assumption.

Residuals Plots against Time

If the data were gathered through time (the data in our example were not) a time plot of the residuals should be done as discussed in Chapter 2. Ideally, this should look like random scatter. Any obvious patterns (lines, curves, fans, etc) indicate time is an important factor and the model which was fit is not adequate to fully describe the relationship. This generally means a multiple regression is needed to explain the response variable.

Identifying Influential Observations

Remember, the two points on the far right side of the original plot looked unusual. Points far away from the center of the range of the predictor variable can be influential; that is, they may have a significant impact on the slope, especially if they do not follow the pattern of the rest of the data. Even if they do not impact the slope, they will cause r and r^2 to be larger than the rest of the data would warrant. To decide if points are influential, delete the suspects, and reanalyze the data.

Here are the first few data values after having deleted the data for the Ford Excursion (310, 10) and GMC Yukon (285, 13).

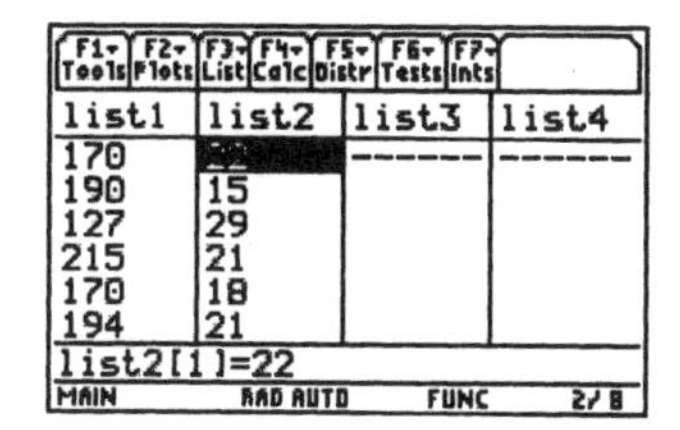

Redrawing the scatterplot (`list1` as `x` and `list2` as `y`) with F5 gives the plot at right. This looks much less linear than the original data scatterplot. One could conceivably think this is no longer "straight enough" for a linear regression. But let's try it anyway.

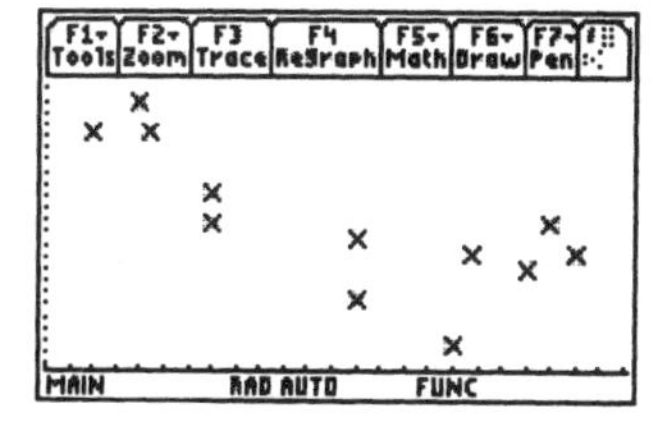

The new linear regression output is at right. The new regression equation is $\textit{Mileage} = 39.39 - 0.10 * \textit{horsepower}$. The slope changed only by about 10%, which is not much. Notice that r and r^2 are much less than before (as expected).

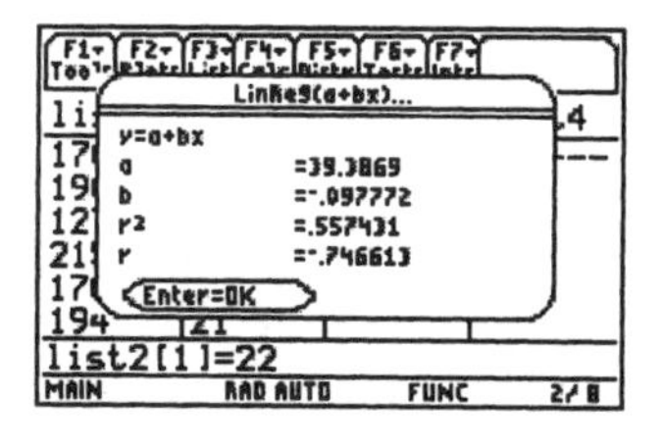

Here's a residuals plot against horsepower (X). We can definitely see the curve in these residuals. They're positive at both ends and negative in the middle.

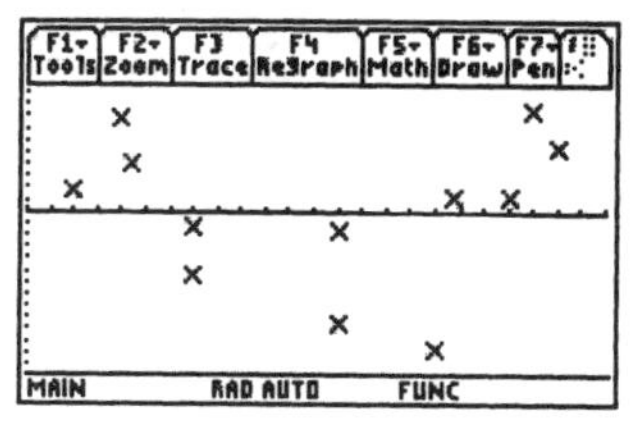

We are left with the following indications: The Ford Excursion and GMC Yukon were influential in this regression; not so much for their impact on the regression equation, but because they made the data much more linear. With these two vehicles removed from the data set, the indication is that a line is *not* the proper model to describe the relationship between horsepower and gas mileage. What is correct? That is beyond the scope of this book; perhaps a model with an x^2 or x^3 term will be better. The TI-89 can calculate these regressions; if you wish they can be checked for adequacy by residuals plots just as we have done here. (It turns out that adding horsepower^2 makes a pretty good relationship.)

Transforming Data

There are two reasons to transform data in a regression setting: to straighten a curved relationship and to transform variability so it is constant around the line. In a single variable case, transformations can be used to make skewed distributions look more symmetric; in the case of a single variable observed for several groups, a transformation can make the different groups look more equally spread.

The table below shows stopping distances in feet for a car tested three times at each of 5 speeds. We hope to create a model that predicts stopping distance from the speed of the car.

Speed (mph)	Stopping Distance (ft)
20	64, 62, 59
30	114, 118, 105
40	153, 171, 165
50	231, 203, 238
60	317, 321, 276

A plot of the data is at right. It looks fairly linear, but it is clear that the stopping distances become more variable with faster speed.

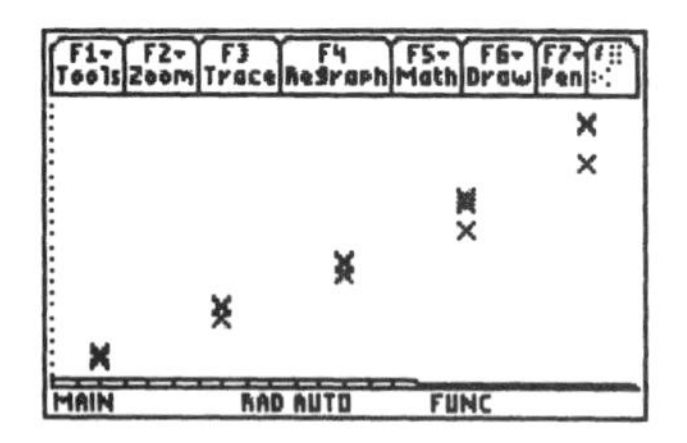

Regression gives the fitted model as $Stoppingfeet = -65.933 + 5.98 * Speed$. The residuals plot against speed (at right) clearly indicates the variability gets larger for faster speeds; it also indicates the true relationship is not linear but curved. Clearly, a transform is indicated – one which will straighten the plot as well as possibly decrease variation.

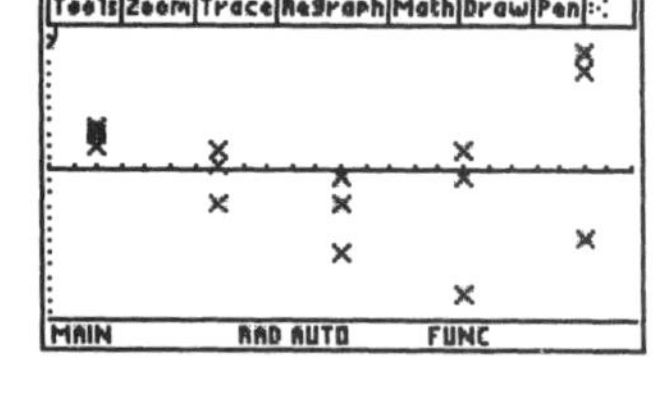

Since the residuals indicate a curve (possibly quadratic), using the square root of stopping distance makes sense. With stopping distance in list2, we need to find the square root of each distance. With one command we can do this, storing the result in a new list, say list3. From the home screen, press [2nd][×] ($\sqrt{\ }$) then [2nd][−] for VAR-LINK, then arrow to list2 and select it, then [)] to close the parentheses, [STO▸]then go back to VAR-LINK and select list3, followed by [ENTER]. The command and result are at right. We see the first few values. To see the entire list, go to the Stat editor.

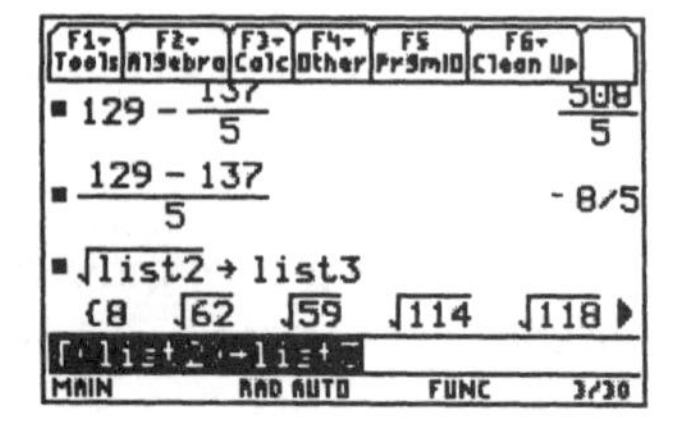

The new scatterplot is at right. The new regression equation is *sqrt(StopDistance)* = 3.303 + 0.235**Speed*. We have $r = 0.9922$, an extremely strong linear relationship. What about a residuals plot?

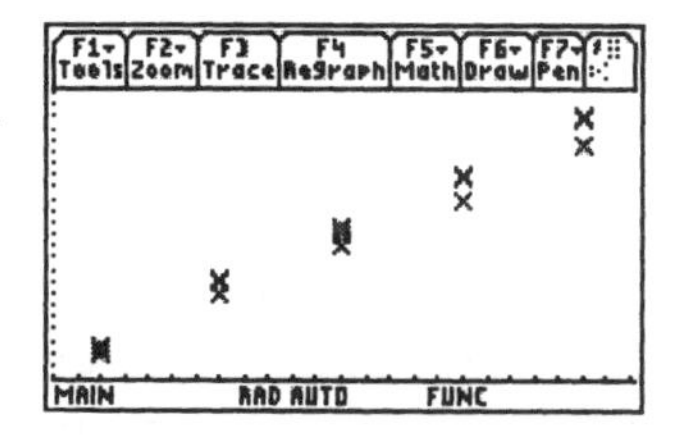

Here is the residuals plot. It's not perfect; the variation still increases with larger values of speed, but is much better than before.
Sometimes there is no "perfect" transform.

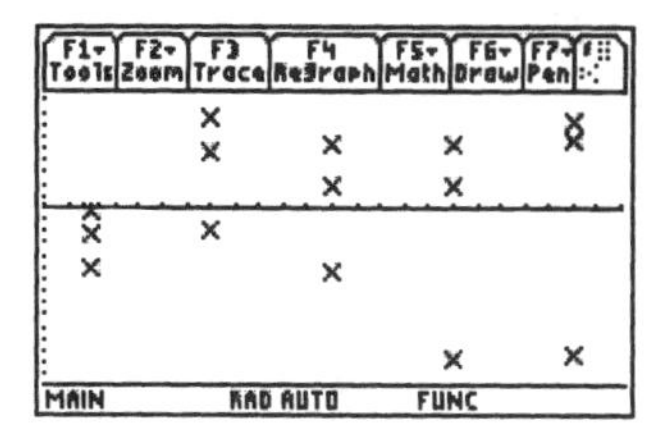

What can go wrong?

What's Dimension Mismatch?

We've seen this one before. Press ESC to quit. This error means the two lists referenced (either in a plot or a regression command) are not the same length. Go to the Stat editor and fix the problem.

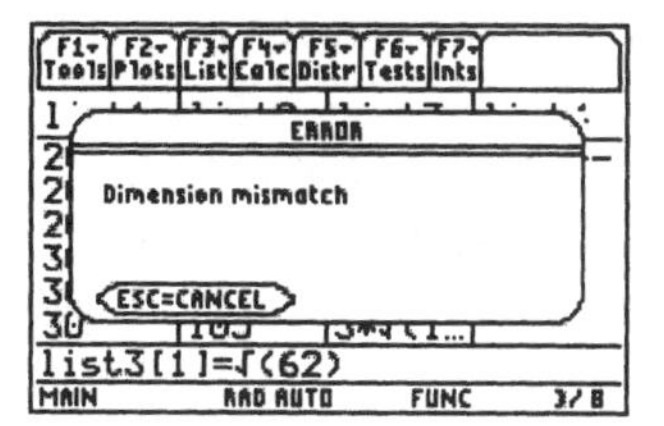

What's that weird line?

This error can come either in a data plot (an old line still resides in the Y= screen) or the stored regression line is showing in the residuals plot, as shown here. The regression line is not part of the residuals plot and shows only because the calculator tries to graph everything it knows about. Press Y= followed by CLEAR to erase the equations, then redraw the graph by pressing GRAPH.

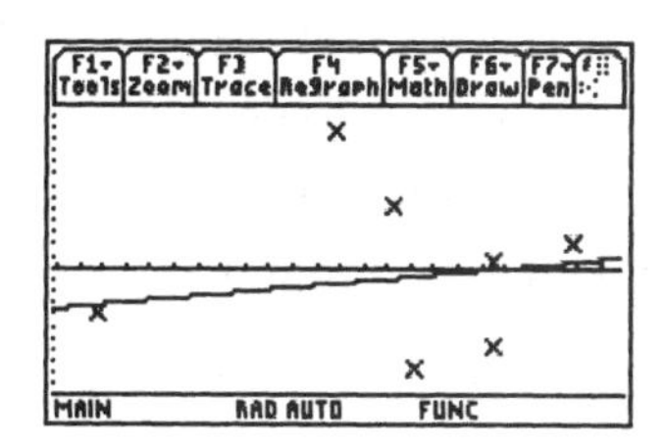

Nonreal result?

This error comes from trying to take the square root (or log) of a negative number. Sorry, can't be done in the real number system. These transforms do not work for negative values. Try something else.

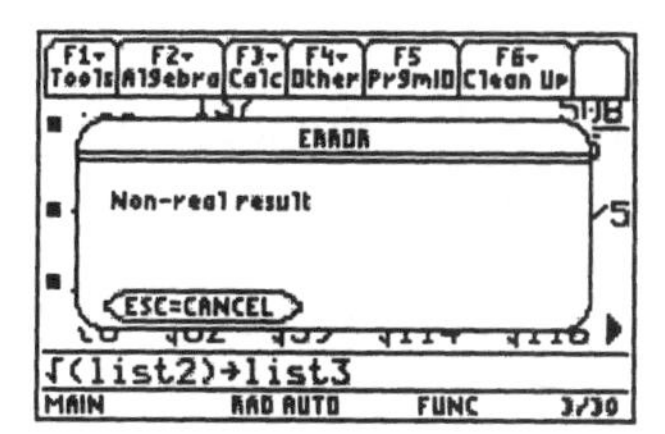

Chapter 19 – Random Numbers

Randomness is something most people seem to have an intuitive sense about. But truly random values are surprisingly hard to get. In fact, calculators (and computers) can't generate true random numbers since any values they obtain are based on an algorithm (that is, a program). But they do give good *pseudorandom* numbers. These have many applications from simulation to selecting samples and assigning treatments in an experiment.

Simulations

Simulations are used to mimic a real situation such as this. Suppose a cereal manufacturer puts pictures of famous athletes in boxes of their cereal as a marketing ploy. They announce that 20% of the boxes contain a picture of Tiger Woods, 30% a picture of Lance Armstrong, and the rest have a picture of Serena Williams. How many boxes of the cereal do you expect to have to buy in order to get a complete set?

You could go out and buy lots of cereal, but that might be expensive. We'll model the situation using random numbers, assuming the pictures really are randomly placed in the cereal boxes, and distributed randomly to stores across the country.

We'll use random digits to represent getting the pictures: Since 20% have Tiger's picture, we'll let the digits 1 and 2 represent getting his picture. Similarly, we'll use digits 3, 4, and 5 (30% of the numbers 1 through 10) to represent getting Lance's picture. The rest (6 through 10) will mean we got a picture of Serena.

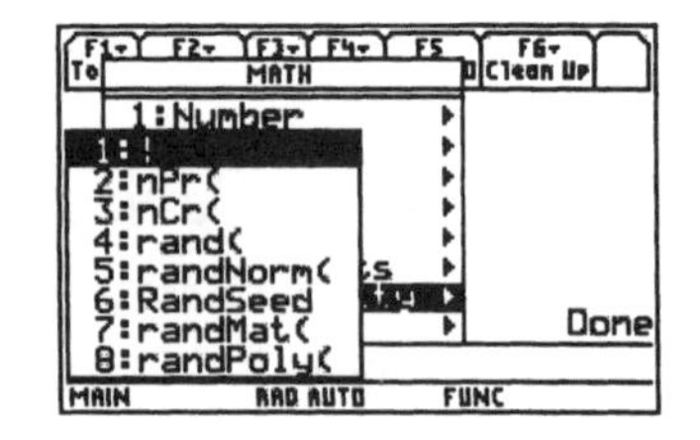

We need to get the random numbers. From the home screen, press [2nd][5] (MATH), then arrow to 7:Probability. Pressing the right arrow displays the menu at right. We want choice 4:rand(. Either arrow to it and press [ENTER] or simply press [4]. The command shell will be transferred to the home screen. Now you need to tell the calculator the boundary values you want. Since we want numbers between 1 and 10, we will enter [1][0][)] . Pressing [ENTER] will get the first random number.

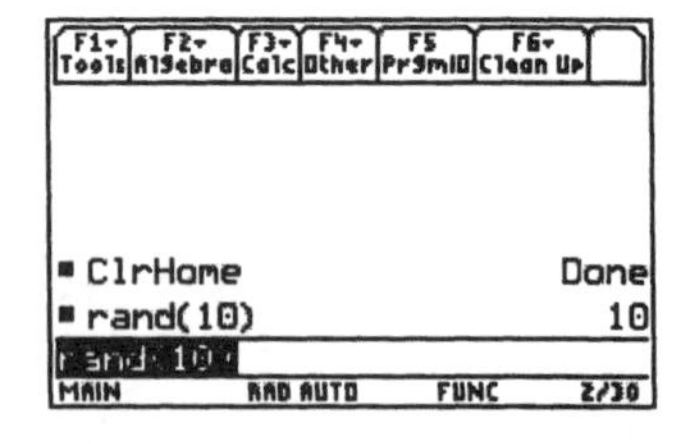

Here is our first random digit: a 10. That means the first box had a picture of Serena. We can continue pressing [ENTER] and get more random digits.

In general, for random integers between 1 and N, the parameter for the rand command is N.

Controlling the sequence of Random Digits

You didn't get the same random number? Not surprising. Random number generation on computers and calculators works from something called a *seed.* In the case of TI calculators, every command you use changes the seed. If a value is explicitly stored as the seed *immediately before* a random number command, the sequence of random digits will be the same every time.

To store a seed, go to the Math, Probability menu as before, but select choice 6:RandSeed. Type in the desired seed value and press [ENTER]. Here, the seed value is 1587. Press [CLEAR] to erase the command from the entry area.

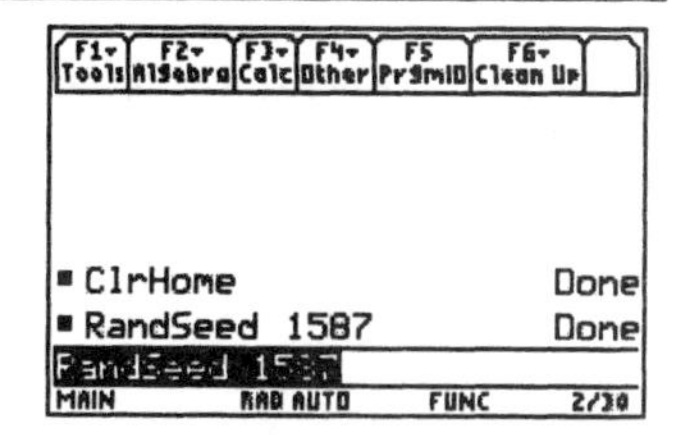

Following this with the same random integer command as before, we get values as shown at right. Yours should be the same. Look at the first five digits. These correspond to (in our example above) getting Lance, Lance, Serena, Serena, and Serena. Even after the fifth "box" we haven't gotten all three pictures. In fact, it takes two more boxes (another Serena then finally a Tiger) for a total of seven boxes to get the complete set.

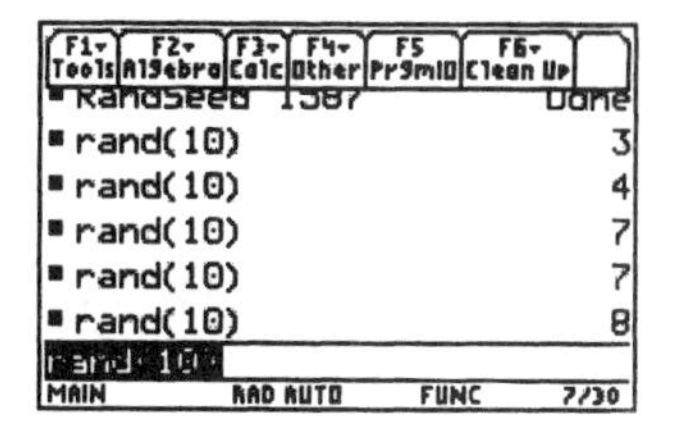

Seeds can also be set using from in the Statistics Editor. Press [F4] (Calc), then 4: Probability, then A:Randseed.

One simulation is not a very good representation. We'd like to know how many boxes it would take to get all three, *on average*. We need to repeat the simulation many times, and take the average value from the many simulations. We could just keep pressing [ENTER] until we've done enough, or we can get many random numbers at once and store them into a list.

Here, we've reset the seed to 18763. Inside the Statistics Editor, move the cursor to highlight the name of a list. Press [F4] then [4] for Probability and select choice 5:randInt(. The parameters are the low number to be generated, the high end of the desired numbers, and how many. End the command by closing the parentheses. Since it could possibly take many boxes of cereal to get all three pictures, we've chosen to store 200 numbers into list1. Pressing [ENTER] executes the command and fills in list1.

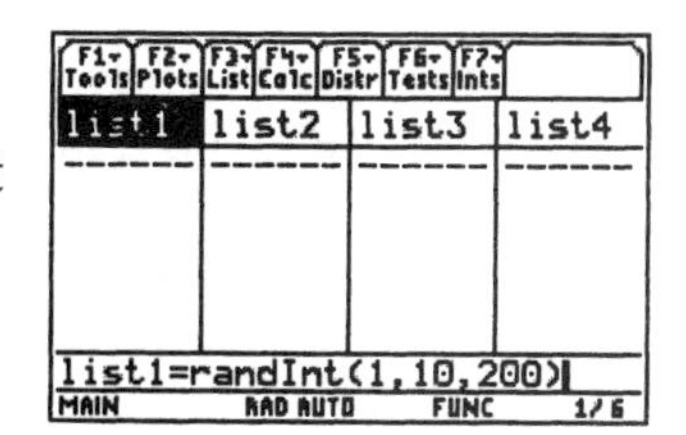

Looking at the list in the editor, the first four digits are 10, 3, 5, and 2. That corresponds to a Serena, Lance, Lance, and Tiger. That trial took four boxes to get the full set. One can continue down the list until several complete sets have been found; then compute the average for all trials as the estimate of the average number of boxes.

Random Normal Data

These calculators can also simulate observations from normal populations in a manner similar to the examples above. On the TI-89 this can either be done from the MATH menu ([2nd][5]) on the home screen to generate single values at a time or from the Statistics/List Editor under the Calc, Probability menu.

Using the command from the Statistics package allows generation of many values into a list as can be done with the randInt command. The command is choice 6:randNorm(from the Calc, Probability menu. The parameters are the mean and standard deviation, and how many numbers to generate. If no value is specified for how many, a single number will be given.

This example models the following: A tire manufacturer believes that the tread life of their snow tires can be described by a Normal model with mean 32,000 miles and standard deviation 2500 miles. You buy 4 of these tires, hoping to drive them at least 30,000 miles. Estimate the chances that all four last at least that long. We will generate output for one trial – a set of four tires.

In this trial, all 4 lasted over 30,000 miles. To further estimate the chance that all four last over 30,000 miles, obtain more repetitions of sets of four tires.

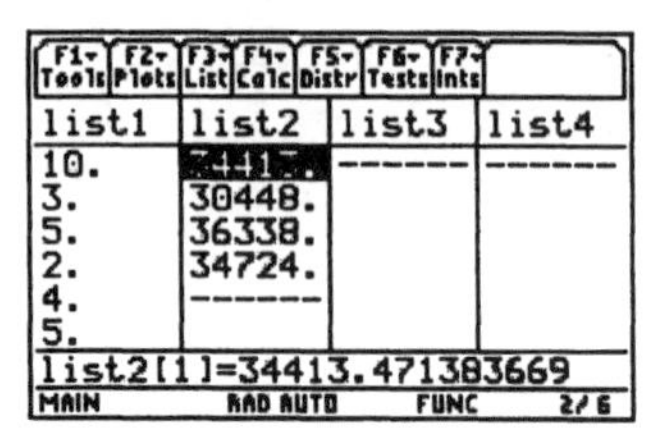

Sampling and Treatment Assignments

Random numbers are the best method for (randomly!) selecting items or individuals to be sampled or treatments to be assigned in an experiment. In the sampling frame (a list of members of the population) number the individuals from 1 to *N*, where *N* is the total number in the list. Use the Random integer command to select those to be sampled. In the case of assigning treatments, if there are, for example two treatments, use random integers to assign half the experimental units to treatment A; the rest will get treatment B.

Chapter 20 – Probability Models

We've already used the calculator to find probabilities based on normal models. There are many more models which are useful. This chapter explores two such models.

Many types of random variables are based on Bernoulli trials experiments. These involve independent trials, only two outcomes possible, and a constant probability of success called p. Two of the more common of these variables have either a Geometric or Binomial model.

Geometric Models

The Geometric probability model is used to find the chance the first success occurs on the nth trial. If the first success is on the nth trial, it was preceded by n-1 failures. Because trials are independent we can multiply the probabilities of failure and success on each trial so we have $P(X = n) = (1-p)^{n-1}p$ which is sometimes written as $P(X = n) = (q)^{n-1}p$. This is generally easy enough to find explicitly, but the calculator has a built-in function to find this quantity as well as the probability the first success comes somewhere on or before the nth trial.

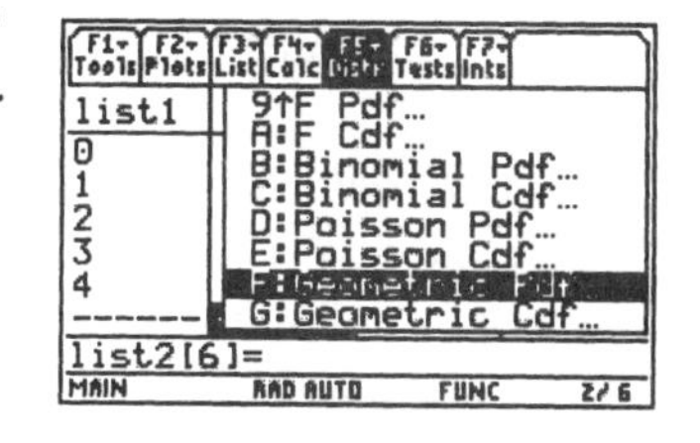

Suppose we are interested in finding blood donors with O-negative blood; these are called "universal donors." Only about 6% of people have O-negative blood. In testing a group of people, what is the probability the first O-negative person is found on the 4th person tested? We want $P(X = 4)$. From the Statistics List Editor, press [F5] (Distr). We want menu choice F:Geometric Pdf. To find the menu option, you can press the down arrow until you find it, press [ALPHA][I] (F), or press the up arrow twice. The last is probably the easiest. Press [ENTER] to select the option.

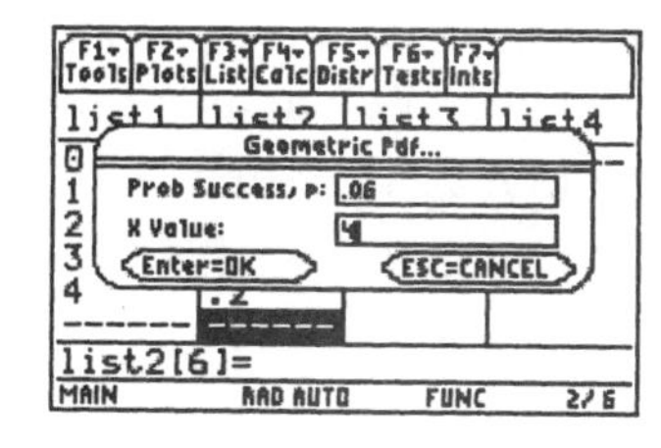

Enter the two parameters for the command: p and x (n). Here, p=.06 and n = 4. Press [ENTER] to find the result. We see there is about a 5% chance to find the first O-negative person on the 4th person tested (assuming of course that the individuals being tested are independent of each other.)

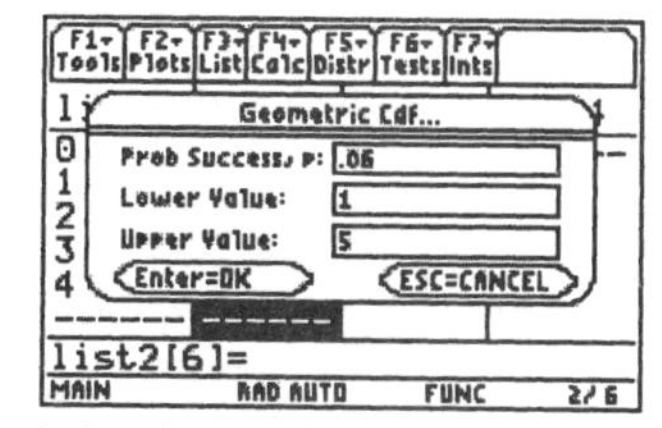

There are some other related questions that can be asked. What is the probability the first O-negative person will be found somewhere in the first 5 persons tested? We want to know $P(X \le 5)$. We could find all the individual probabilities for 1, 2, 3, 4, and 5 and add them together but there is an easier way. We really want to "accumulate" all those probabilities into one, or find the *cumulative* probability. This uses menu choice G:Geometric Cdf. The parameters are a little different. We first input the probability of a success (here, .06); then the low end of interest (1) and finally the high end of interest (5). After pressing [ENTER] we see that finding the first O-negative person within the first 5 people tested should happen about 26.6% of the time.

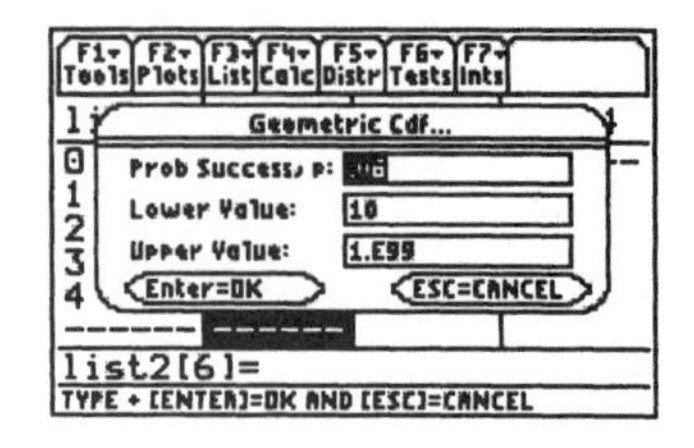

What's the chance we'll have to test at least 10 before we find an O-negative person? We want $P(X \geq 10)$. Since there could (possibly) be an infinite number of people tested to find the first O-negative person, we will use a low end of 10 and an upper end of ∞ ([1][EE][9][9]). After pressing [ENTER], we see the chance we'll have to test at least 10 people to find the first O-negative person is 57.3%.

Binomial Models

Binomial models are interested in the chance of k successes occurring when there are a fixed number (n) of Bernoulli trials.

Consider a family planning on five children. What is the chance they will have three girls? Girls and boys do not actually each happen half the time. Boys are really born about 51.7% of the time. This means girls are born 48.3% of the time. We know pregnancies are independent of each other (unless multiple births are involved). In order to have three girls, the family could have girl, girl, boy, girl, boy or boy, boy, girl, girl, girl, or any one of several different arrangements, each of which will have the same probability. How many arrangements are possible so that there are 3 girls in 5 children? The Binomial coefficient provides the answer to this question. The coefficient itself is variously written as $\binom{n}{k}$ or nCk and is read as "n choose k." From the home screen, then press [2nd][5] (`Math`) and arrow to `7:Probability`. Press the right arrow to get the submenu and select choice `3:nCr(`. The command is transferred to the input portion of the home screen. Enter n (5 here) a comma and k (3 here). Finish the command by closing the parentheses and press [ENTER].

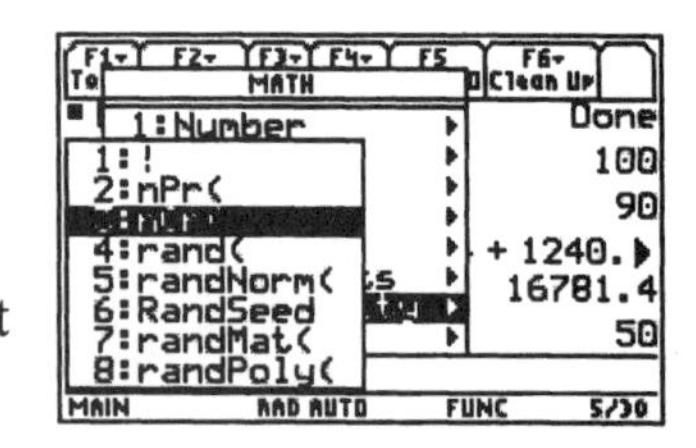

We find there are 10 possible arrangements of three girls in a family of five children. We can then find the probability of three girls in a family of five children as $10(.517)^2(.483)^3 = 0.301$. This means about 30% of families with five children should have three girls.

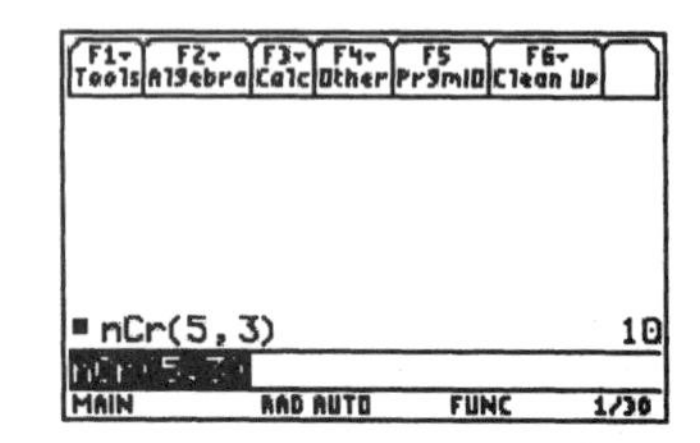

There is an easier way to find these binomial probabilities. Looking back at the portion of the `Distr` menu shown on the previous page there are two menu options that will help us here: `B:Binomial Pdf` and `C:Binomial Cdf`. Just as in the case of the geometric model discussed above, the pdf menu choice gives $P(X = k)$ and the cdf gives probabilities for specified low and high values of interest. Three examples follow.

From the Statistics List editor, Press [F5] (`Distr`). Either arrow down to choice B and press [ENTER], press [alpha][(] (B), or press the up arrow until you find choice B and press [ENTER]. On the input screen there are three parameters to enter: n (the number of trials), p (the probability of success), and x (the value of interest).

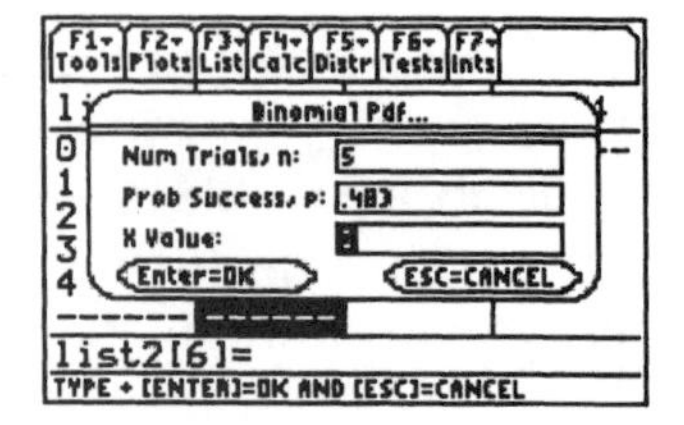

Here we have entered the command and pressed [ENTER] to find the result. This is just the quantity we calculated explicitly above, namely there is about a 30% chance that a family with five children will have three girls.

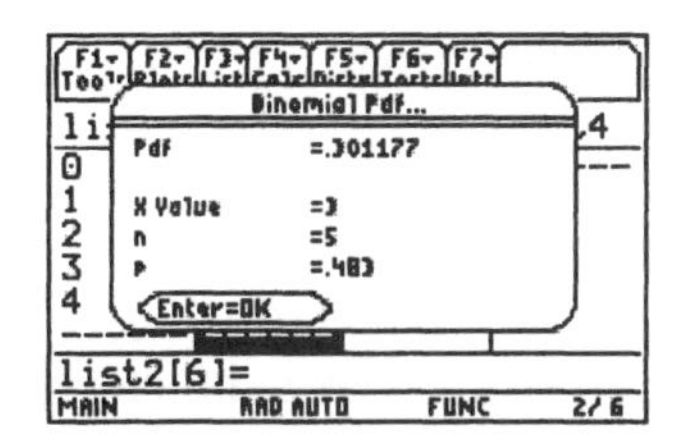

Returning to the prior example about blood donors, what is the probability that if 20 donors come to the blood drive, there will be 3 O-negative donors? From the screen at right, we see this is 8.6%.

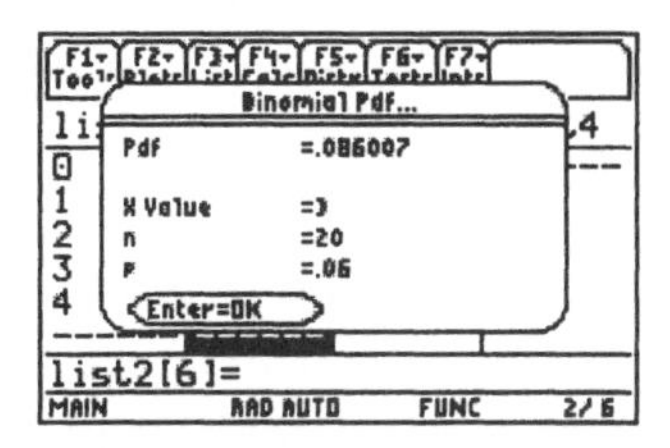

As with geometric models, we can find the use the cdf command to find cumulative probabilities. Remember, the calculator adds individual terms for the values of interest specified. For instance, what is the chance of at most 3 O-negatives in a blood drive with 20 donors? We see this is very likely to happen. We'll expect 3 or fewer O-negative donors in 20 people about 97% of the time.

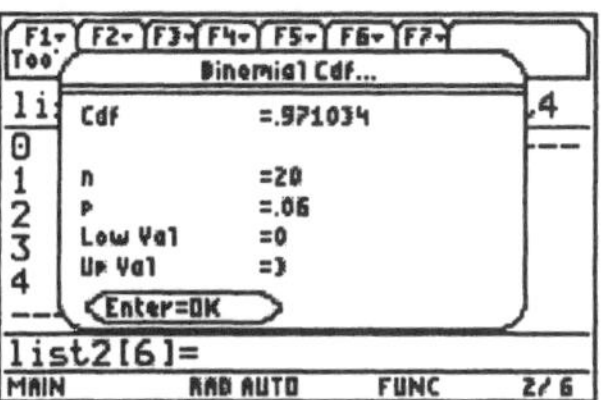

What's the chance there would be more than 2 O-negative donors in a group of 20? Again, since we are looking for P($X > 2$) we want everything above 2, so the low end of interest is 3. The high end is 20. In a group of 20 donors, we'll get more than 2 O-negative people about 11.5% of the time.

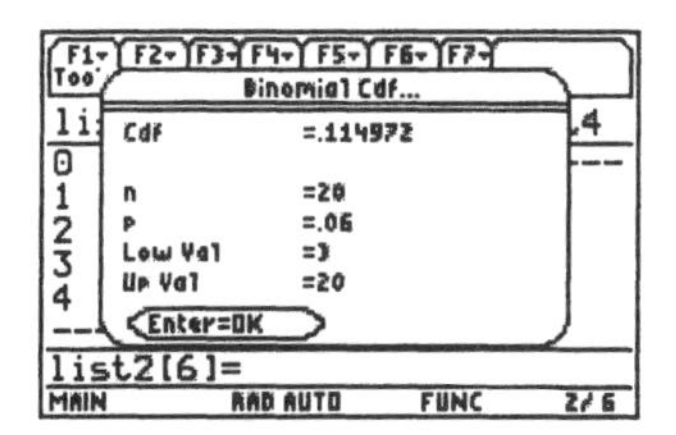

A final word about binomials. The TI-89 (and most computer applications as well) cannot deal with large values of n. This is because the binomial coefficient becomes too large very quickly. However, when n and p are sufficiently large (generally, both of $np \geq 10$ and $n(1-p) \geq 10$ must be true to move the distribution away from the ends so it can become symmetric) binomials can be approximated with a normal model. One uses the normalcdf command described in Chapter 4 specifying the mean as the mean of the binomial ($\mu = np$) and the standard deviation as that of the binomial ($\sigma = \sqrt{np(1-p)}$).

Suppose the Red Cross anticipates the need for at lease 1850 units of O-negative blood this year. They anticipate having about 32,000 donors. What is the chance they will not get enough O-negative blood? We desire P($X < 1850$). We have calculated the mean to be 1920 and the standard deviation to be 42.483. Practically speaking the low end of interest is 0, but remember the normal distribution extends to $-\infty$. It appears there is about a 5% chance there will not be enough O-negative in the scenario discussed.

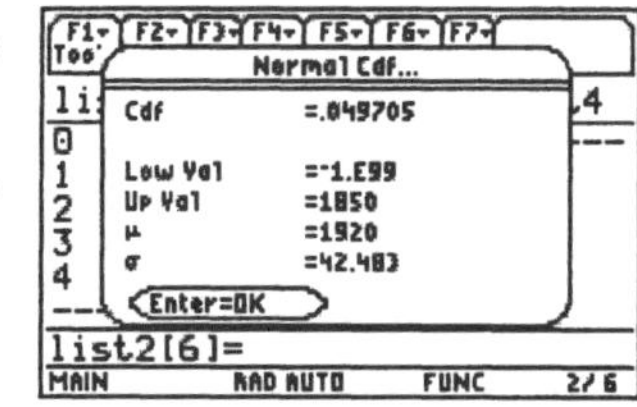

What can go wrong?

Domain error?

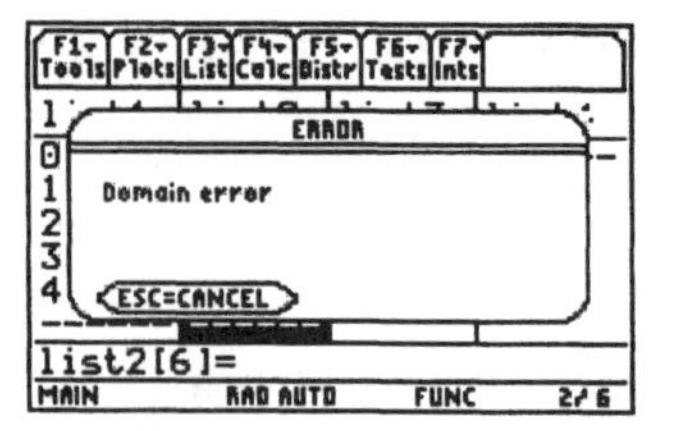

This error is normally caused in these types of problems by specifying a probability as a number greater than 1 (in percent possibly instead of a decimal) or a value for n or x which is not an integer. Reenter the command giving p in decimal form. Pressing [ESC] will return you to the input screen to correct the error. This will also occur if n is too large in a binomial calculation; if that is the case, you need to use the normal approximation.

How can the probability be more than 1?

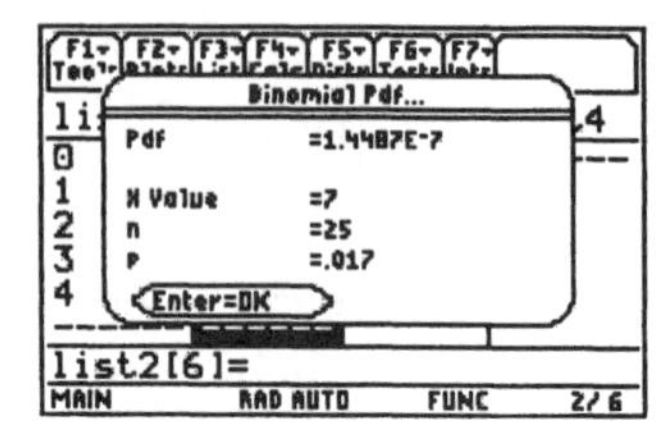

It can't. As we've said before, if it looks more than 1 on the first glance, check the right hand side. This value is 1.45×10^{-7} or 0.0000001, not 1.45.

Chapter 21 – Inference for Proportions

We know the sample proportion, $\hat{p}$, is normally distributed if both $np \geq 10$ and $n(1-p) \geq 10$. With this fact we can obtain probabilities based on the normal model of obtaining certain sample proportions. Inference asks a different question. Based on a sample, what can we say about the true population proportion? Confidence intervals give ranges of believable results along with a probability statement giving our level of certainty that the interval contains the true value. Hypothesis tests are used to decide if a claimed value is or is not reasonable based on the sample.

Confidence Intervals for a single proportion

Sea fans in the Caribbean Sea have been under attack by a disease called *aspergillosis*. Sea fans which can take up to 40 years to grow can be killed quickly by this disease. In June 2000, members of a team from Dr. Drew Harvell's lab sampled sea fans at Las Redes Reef in Akumai, Mexico at a depth of 40 feet. They found that 54 of the 104 fans sampled were infected with the disease. What might this say about the prevalence of the disease in general? The observed proportion, $\hat{p} = 54/104 = 51.9\%$ is a point estimate of the true proportion, p. Other samples will surely give different results.

We can use the calculator to obtain a confidence interval for the true proportion of infected sea fans. From the Statistics list editor press [2nd][F2](F7 : Ints). There are several types of confidence intervals that can be created. We'll talk about most of them later. The interval we want here is choice 5:1-PropZInt. Either arrow to it and press [ENTER] or press [5].

Here is the input screen. Simply enter the number of observed "successes," which here is the number of infected sea fans, 54, press the down arrow, then enter the number of trials (the 104 fans observed), press the down arrow again to enter the desired level of confidence, finally press the down arrow again and press [ENTER] to calculate the results.

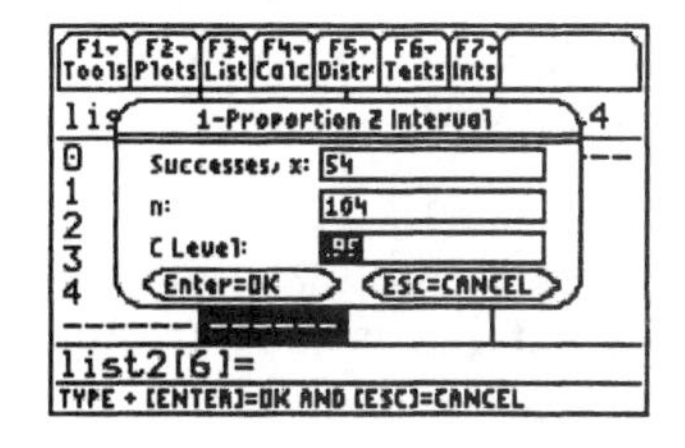

How much confidence? That is up to the individual researcher. The trade-off is that more confidence requires a wider interval (more possible values for the parameter). 95% is a typical value, but the level is generally specified in each problem.

Here are the results. The interval is 0.423 to 0.615. Remember the calculator tends to give more decimal places than are really reasonable. Usually reporting proportions to tenths of a percent is more than enough. Your instructor may give other rules for where to round final answers. The output also gives the sample proportion and the sample size.

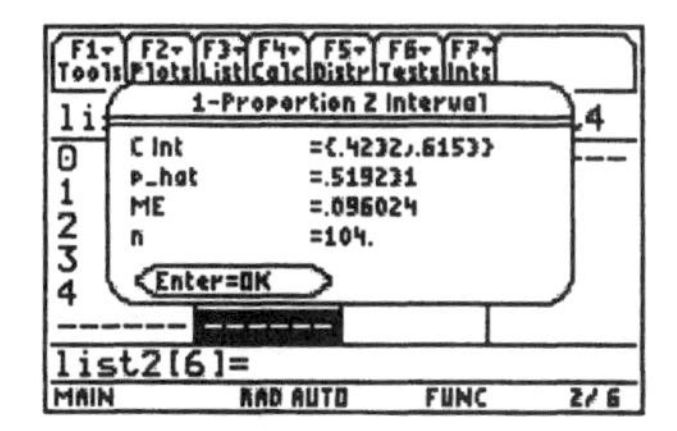

What can we say about the prevalence of disease in sea fans? Based on this sample, we are 95% confident that between 42.3% and 61.5% of Las Redes sea fans are infected by the disease. The estimated proportion from the sample is 51.9% and the margin of error (ME) is 9.6%.

In August 2000, the Gallup poll asked 507 randomly sampled adults the question "Do you think the possession of small amounts of marijuana should be treated as a criminal offense?" Of these, 47% answered "No." What can we conclude from the survey?

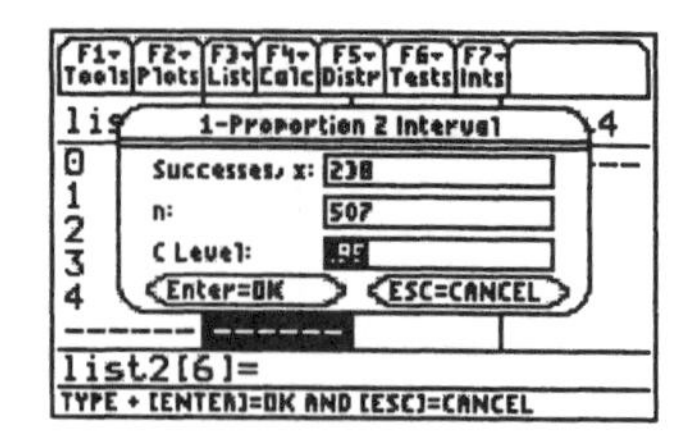

Results from polls are usually given in (rounded) whole percents. In order to create a confidence interval using the calculator we need the number of respondents who answered "No." Multiplying 0.47*507 gives 238.29. Since there can't be a fraction of a response, round this number to 238 "No" responses. The input screen is at right.

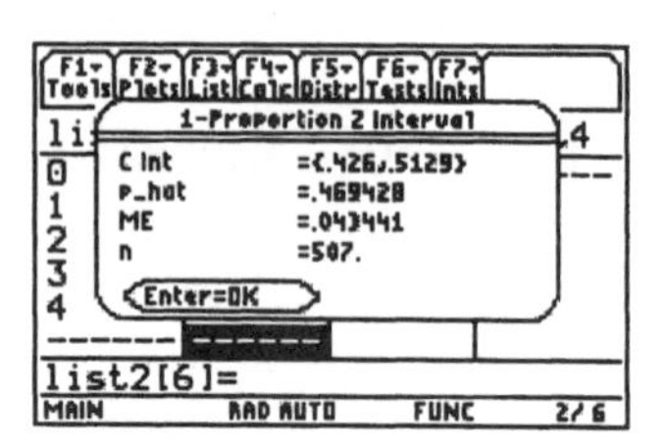

Based on this poll, we can be 95% confident the proportion of Americans who think possession of small amounts of marijuana should not be treated as a criminal offense is between 42.6% and 51.3%.

Hypothesis tests for a single proportion

Confidence intervals give ranges of believable values for the parameter (in this case the proportion of successes). Hypothesis tests assess the believability of a claim about the parameter. Certainly, if a claimed value is contained in a confidence interval it is plausible. If not, it is unreasonable. Formal tests of hypotheses assess the question somewhat differently. The results given include a test statistic (here a z-value based on the standard model) and a p-value. The p-value is the probability of a sample result as or more extreme than that actually obtained, given the claimed value of the parameter. Large p-values argue in support of the claim, small ones argue against it; in essence, if the claimed value were true the likelihood of observing what was seen in the sample is very small.

In some cultures, male children are valued more highly than females. In some countries with the advent of prenatal tests such as ultrasound, there is a fear that some parents will not carry pregnancies of girls to term. A study in Punjab India[1] reports that in 1993 in one hospital 56.9% of the 550 live births were males. The authors report a baseline for this region of 51.7% male live births. Is the sample proportion of 56.9% evidence of a change in the percentage of male births?

From the Statistics List Editor, press [2nd][F1] (F6, `Tests`). Just as there are several types of confidence intervals there are several types of tests which can be performed. Select choice `5:1-PropZTest` by either pressing [5] or arrowing to the selection and pressing [ENTER]. We are asked for p_0, the posited value which is 51.7% in this case. Enter the proportion as the decimal 0.517. Then we need the number of "successes" (Multiplying 0.569*550 gives 312.95 which rounds to 313). The number of trials, n, is 550. Then we need a direction for the alternate hypotheses (what we hope to show). This is usually obtained from the form of the question. In this case, we want to know if the proportion has changed which might argue for selecting the $prop \neq p_0$ alternative, but we suspect that this proportion should increase from the baseline if male births are being selected. Press the right arrow to

[1] "Fetal Sex determination in infants in Punjab, India: correlations and implications", E.E. Booth, M. Verna, R. S. Beri, *BMJ*, 1994; 309:1259-1261 (12 November).

see the choices for the alternative, arrow down to highlight prop > p0 and press [ENTER] to select that alternative. Finally, there are two choices for output. Selecting Calculate merely gives the results. Selecting Draw draws the normal curve and shades in the area corresponding to the p-value of the test. The input screen for the test is at right.

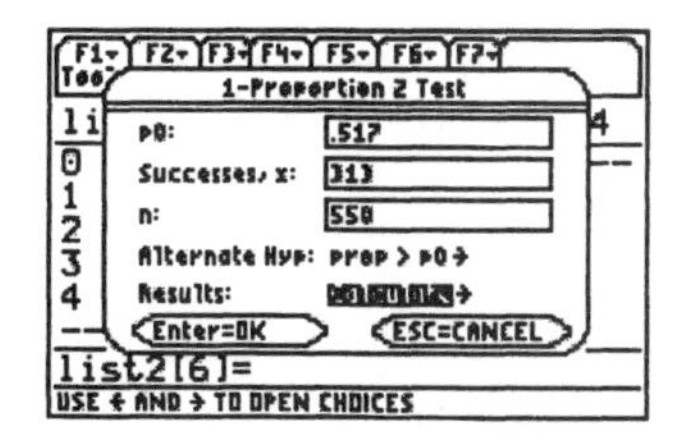

At right is the output from selecting Calculate. The first line of the output gives what the calculator understood the alternate hypothesis to be. Always check that this is what you intended, as it can make a difference in the p-value. The value of the test statistic is $z = 2.44$. This means that if there if there were still 51.7% male births, the observed 56.9% is 2.44 standard deviations above the mean. The p-value for the test is 0.0072. This means that if the proportion of male births is still 51.7%, we would observe a value of 56.9% or greater only about 7 times in 1000. Since this is very rare, we will reject the null hypothesis and conclude that we believe, based on these data, the true proportion of male births in Punjab is now greater than the baseline 51.7%.

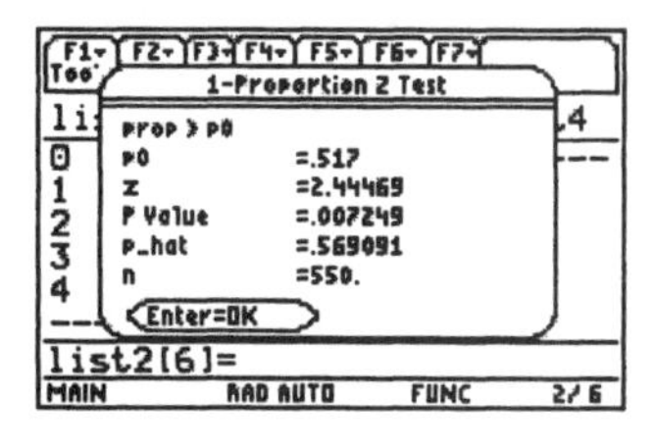

Here is the output when DRAW is selected. There is not as much information given, but the test statistic and p-value are reported. Since this p-value is so small, not much is shown as shaded.

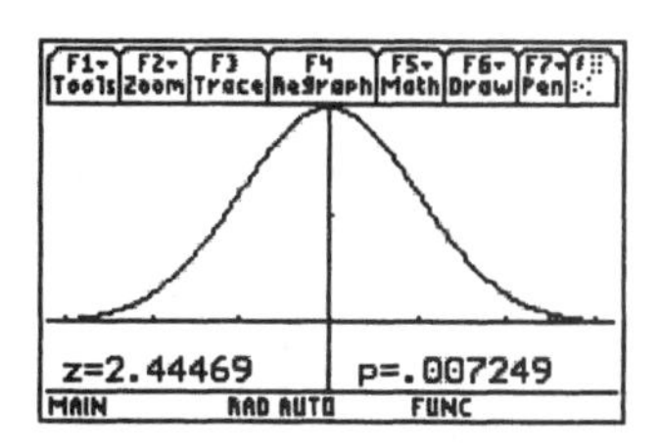

When a test rejects the null hypothesis as in the example above (remember, we decided the proportion of male births is now more than the baseline) it is good practice to report a confidence interval for where the parameter is, based on the sample. Find a 95% confidence interval for the true proportion of male births in Punjab, as detailed above. Notice that the data entry in the input screen is preserved from the hypothesis test. Based on these data, we are 95% confident the true proportion of male births in Punjab is now between 52.8% and 61.1%.

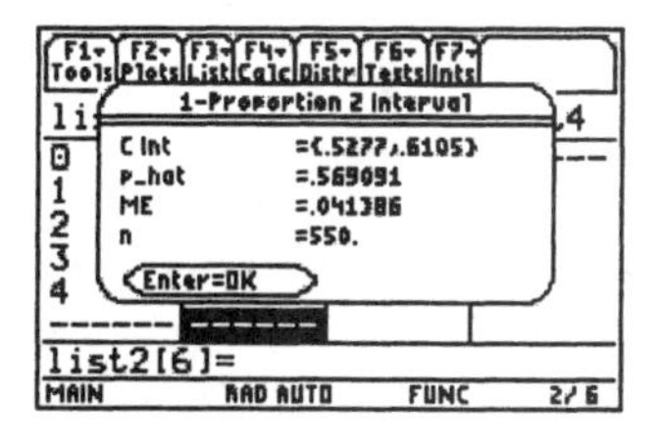

Is there a home field advantage in baseball? In 2002, the home team won 1314 of 2425 games. That's 54.2%. It's more than the 50% we would expect if there were no home field advantage, but is it enough bigger to say it's statistically significant? The input screen is shown at right. Notice that if there is a home team advantage, we'd expect the home team to win more than 50% of games, so the alternate is $> p_0$.

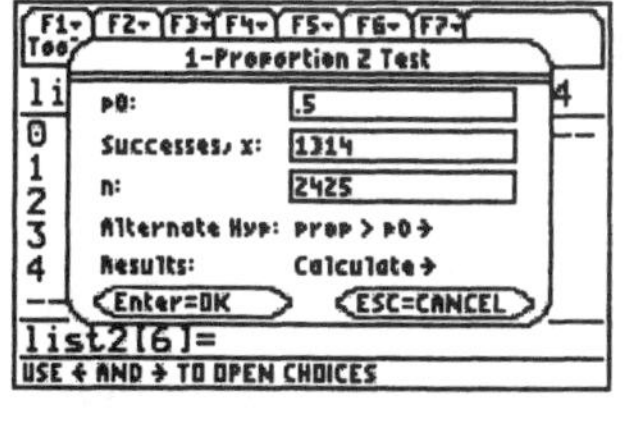

Here are the results of the test. The z-statistic is 4.12; the p-value is 0.000019. This extremely small p-value argues that these data show there is indeed a home field advantage.

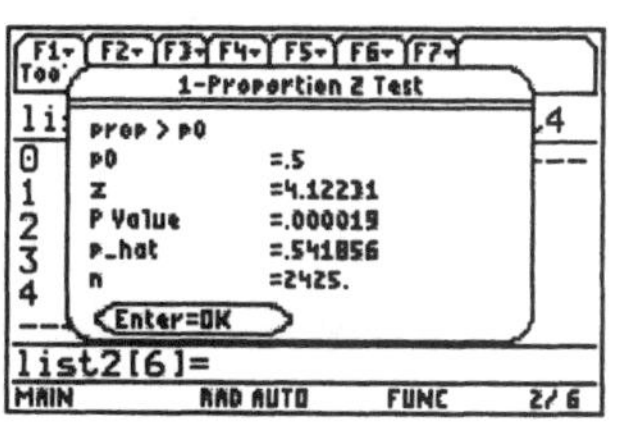

How large is the advantage? Based on these data, we are 95% confident the home team will win between 52.2% and 56.2% of baseball games.

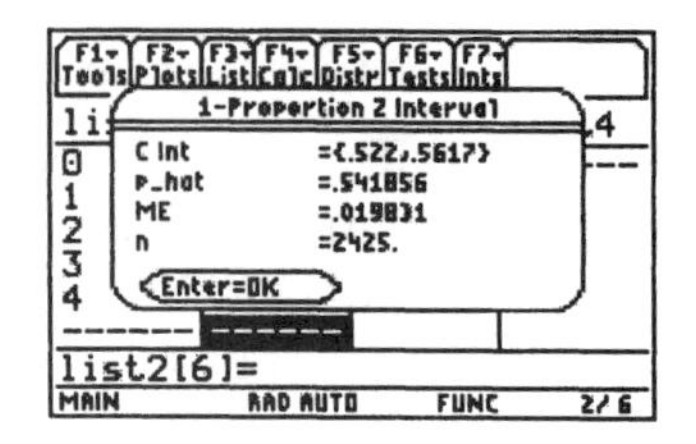

Confidence Intervals for the difference in two proportions

Who are typically more intelligent, men or women? To find out what people think, the Gallup Poll sampled 520 women and 506 men. They showed them a list of attributes and asked them to indicate whether each attribute was "generally more true about men or women."[2] When asked about intelligence, 28% of men thought men were normally more intelligent, but only 14% of women agreed. The difference is 14%, which looks large, but is it large enough to be meaningful or is it due to sampling variability? We will first compute a confidence interval for the difference in the true proportions, $p_M - p_W$. If this interval contains 0, there is no statistical evidence of a difference.

From the Statisics `Ints` menu select choice `6:2-PropZInt...` by either arrowing to it and pressing [ENTER] or by pressing [6]. The calculator uses groups 1 and 2 not men and women and calculates results based on $group1 - group2$. Decide (and keep note of) which group you call which. Since it is usually easier to deal with positive numbers, we will call the males group 1. As before, we calculate numbers of "successes" for men as 0.28*506 = 141.68 (round this to 142) and for women as 0.14*520 = 72.8 (which rounds to 73). The input screen should look like the one at right.

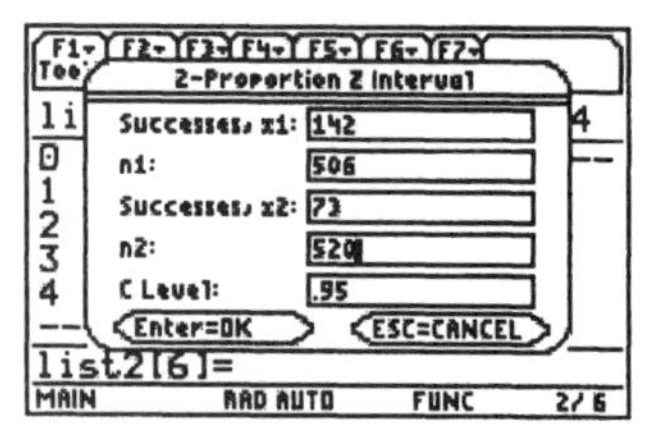

The calculated interval is 0.091 to 0.190. This means we are 95% confident, based on this poll, the proportion of men who think men are more intelligent is between 9.1% and 19.0% more than the proportion of women who think men are more intelligent. Since the interval does not contain 0, there is a definite difference in the two genders.

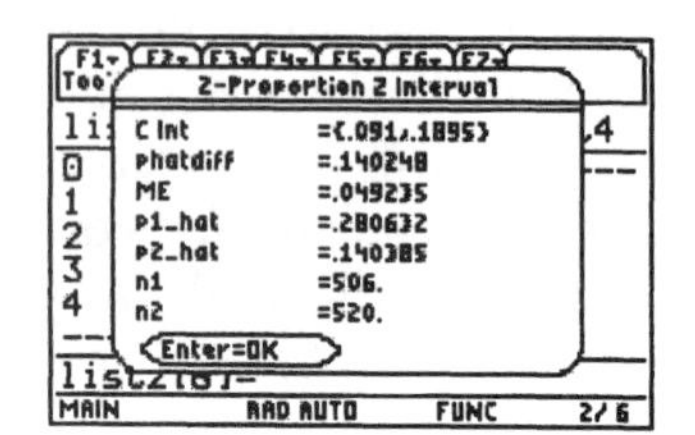

Hypothesis tests for a difference in proportions

The National Sleep Foundation asked a random sample of 1010 U.S. adults questions about their sleep habits. The sample was selected in the Fall of 2001 from random telephone numbers.[3] Of interest to us is the difference in the proportion of snorers by age group. The poll found that 26% of the 184 people age 30 or less reported snoring at least a few nights a week; 39% of the 811 people in the older group reported snoring. Is the observed difference of 13% real or merely due to sampling variation?

The null hypothesis is there is no difference, or $p_1 - p_2 = 0$. (The calculator and most computer statistics packages can only test assumed differences of 0; if there were an assumed difference, say the belief is that older people had 10% more snorers than young people, one would need to compute the test statistic "by hand.")

[2] http://www.gallup.com/poll/releases/pr010221.asp

[3] 2002 *Sleep in America Poll*, National Sleep Foundation. Washington D.C.

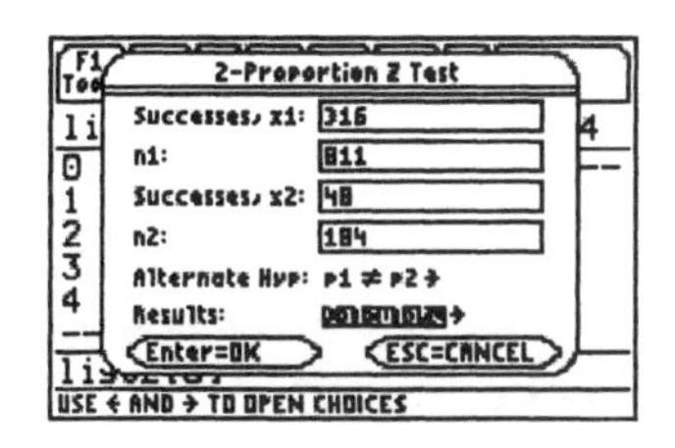

Decide which group will be group1. We will use the older people as group 1. (There will be no difference in the results, but again, positive numbers are generally easier for most people to deal with.) From the STAT, TESTS menu select choice 6 : 2-PropZTest. The number of snorers in the older group is $0.39*811 = 316.29$ (rounded to 316); for the younger group the number of snorers is $0.26*184 = 47.84$ which rounds to 48. The chosen alternate is $p1 \neq p2$ since we just want to know if there is a difference.

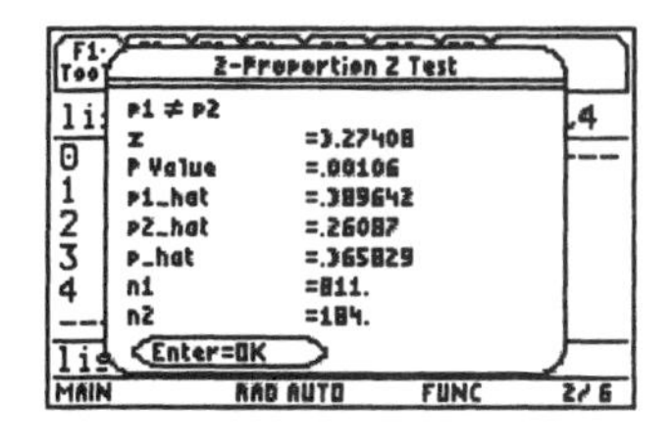

Here are the results. Be careful here, as there are lots of p's floating around. We first see the chosen alternative, $p1 \neq p2$, The value of the test statistic is $z = 3.27$. The p-value for the test is given next: p = 0.0011. The next values given are the observed proportions in each group, $\hat{p}_1$ and $\hat{p}_2$ then an overall $\hat{p}$ which represents the observed proportion, *if there were no difference in the groups*. The sample sizes for the two groups are given as well. Since the p-value for the test is so small, we believe there is a difference in the rate of snorers based on this poll. We can further say the proportion of snorers is greater in older people than in those under 30 since their observed proportion was larger.

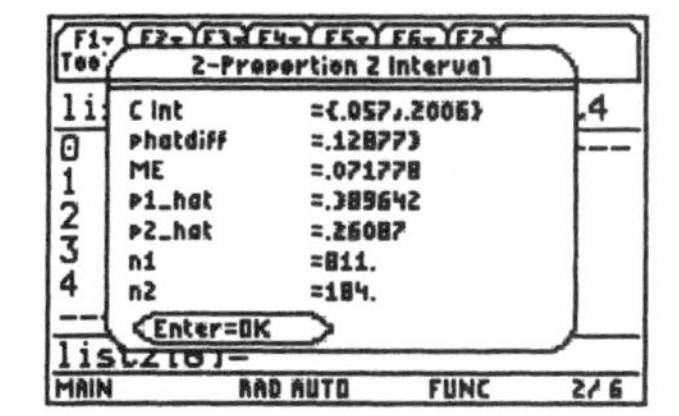

How big is the difference? We are 95% confident, based on this data the proportion of snorers in older adults is between 5.7% and 20.1% larger than for those under 30.

What can go wrong?

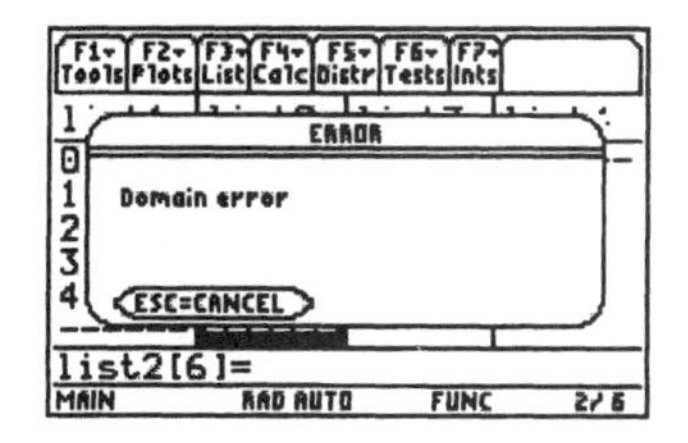

Domain Error?

This error stems from one of two types of problems. Either a proportion was entered in a 1-PropZTest which was not in decimal form or the numbers of trials and/or successes was not an integer. Go back to the input screen and correct the problem.

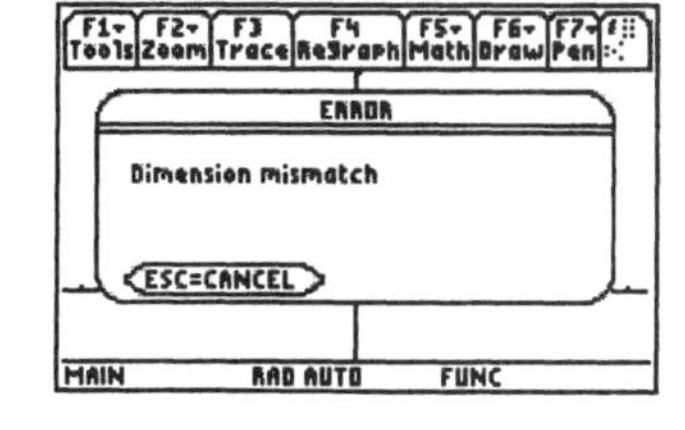

Dimension Mismatch?

This can be caused by selecting the DRAW option if another Statistics plot is turned on. Either go to the STAT PLOT menu ([2nd][Y=]) and turn off the plot or redo the test selecting CALCULATE.

Bad Conclusions.

Small p-values for the test argue against the null hypothesis. If the p-value is small, one rejects the null hypothesis and believes the alternate is true. If the p-value is large, the null hypothesis is not rejected; this does *not* mean it is true – we simply haven't gotten enough evidence to show it's wrong. Be careful when writing conclusions to make them agree with the decision.

Chapter 22 – Inference for Means

Inference for means is a little different than that for proportions. Most introductory statistics texts base this on standard normal models which is truly appropriate only if the population standard deviation, σ, is known. In most cases this is not true; the only time one might really believe σ is known is in the case of quality control sampling where a production line has been tracked for a long time. If σ is not known confidence intervals and hypothesis tests should be based on t distributions. These become the standard normal distribution when the sample size is very large (infinite).

Small sample sizes give rise to their own problems. If the sample size is less than about 30, the Central Limit Theorem does not apply, and one cannot assume the sample mean has a normal distribution. In the case of small samples, you must check that the data come from a (at least approximately) normal population, usually by normal probability plots since histograms are not useful with small samples.

Confidence Intervals for a mean

Residents of a small northeastern town who live on a busy street are concerned over vehicles speeding through their area. The posted speed limit is 30 miles per hour. A concerned citizen spends 15 minutes recording the speeds registered by a radar speed detector that was installed by the police. He obtained the following data:

29	34	34	28	30	29	38	31	29	34	32	31
27	37	29	26	24	34	36	31	34	36	21	

We want to estimate the average speed for all cars in this area, based on the sample. Enter the data in a list. Here, I have entered them into `list1`. This is a small sample – there are only 23 observations, so we should check to see if the data looks approximately normal.

A normal plot of the data looks relatively straight, with no outliers, so it's reasonable to continue. This plot shows some granularity (repeated measurements of the same value) but no overt skewness. If you've forgotten how to create normal probability plots, return to Chapter 17.

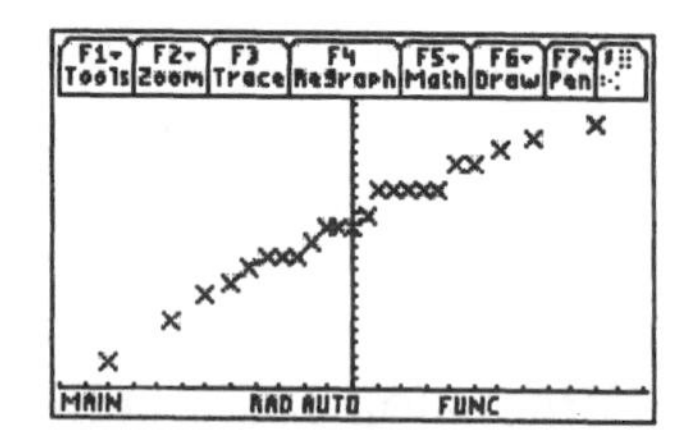

From the Statistics list editor, press [2nd][F2] (F7, `Ints`). Select choice `2:TInterval`. You have two choices for data input: using data in a list such as we have or inputting summary statistics from the sample. Pressing the right arrow allows you to make the selection. Press [ENTER] to get the next input screen. Enter the name of the list with the data ([2nd][-] takes you to the VAR-LINK screen). Each observation occurred once, so `Freq` should be 1. If there were a separate list of frequencies for each data value, that would be entered here. Enter the desired amount of confidence (here, 90%, but in decimal form) and finally press [ENTER] to perform the calculation.

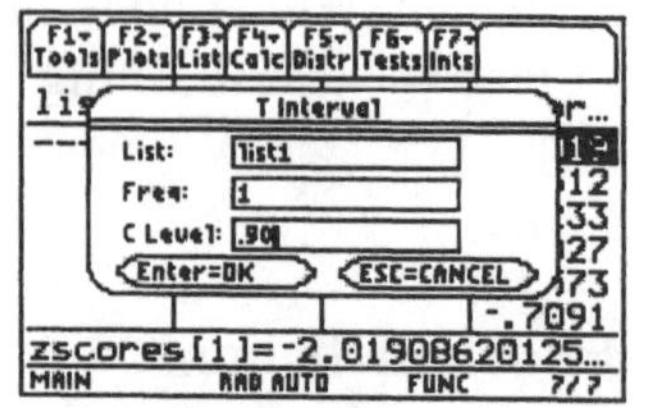

Here are the results. Based on this sample, we are 90% confident the average speed for all cars on this road is between 29.5 and 32.6 miles per hour. There are two caveats here: the first is that this was not a truly random sample but a convenience one (only one 15 minute period was sampled). Also, the presence of the radar speed detector may have influenced the drivers at that time. Drivers may be driving over the posted 30 miles per hour limit, but since 30 is included in the interval, we have not shown that is wrong.

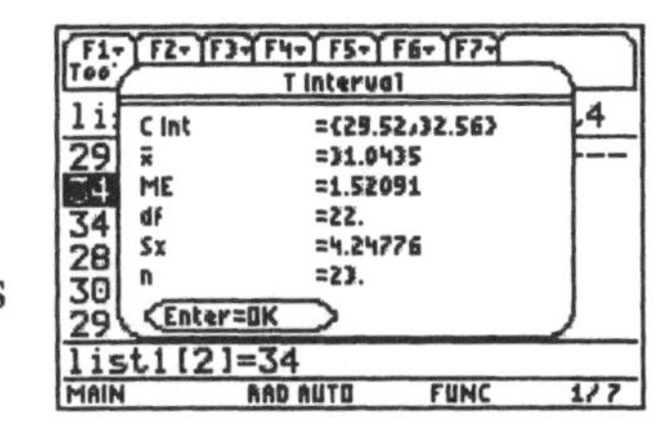

What if we don't have the data? In the case of a small sample size, one must assume the data comes from an approximately normal population. If the sample is "large" the Central Limit Theorem will apply.

A nutrition laboratory tests 40 "reduced sodium" hot dogs, finding that the mean sodium content is 310 mg with a standard deviation of 36 mg. What is a 99% confidence interval for the mean sodium content of this brand of hot dog? Here we have the Data Input Method to be `Stats`. When this is done, the input screen changes to ask for the sample mean, standard deviation, sample size and confidence level.

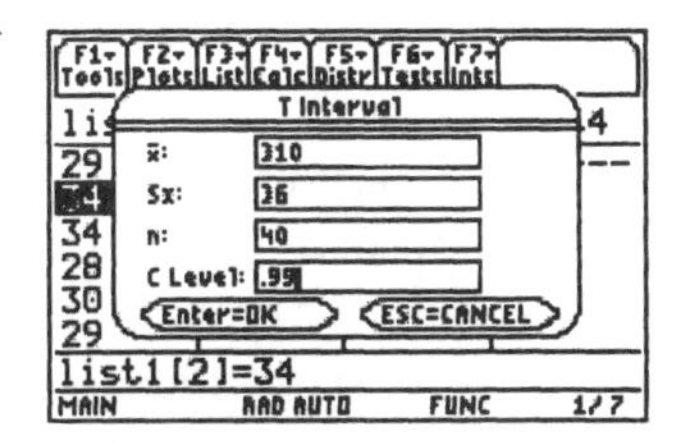

Pressing [ENTER] to calculate the interval tells us we are 99% confident, based on this sample the mean sodium content for these hot dogs is between 294.6 and 325.4 mg.

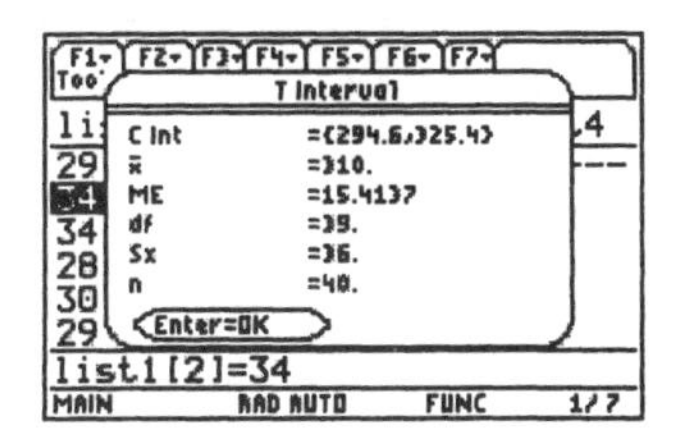

A one sample test for a mean

We can also do a hypothesis test to decide whether the mean speed is more than 30 mph. From the Statistics list editor, press [2nd][F1] (F6, `Tests`) and select choice `2:Ttest`. Again you are asked what the Data Input Method is. We are still using data in `list1`, so this should be `Data`. μ_0 is set to 30 since that's the speed limit we're comparing against. The alternate has been selected as $\mu > \mu_0$ since we want to know if people are going too fast, on average. Notice we have the options of Calculate and Draw here, just as we did on tests of proportions.

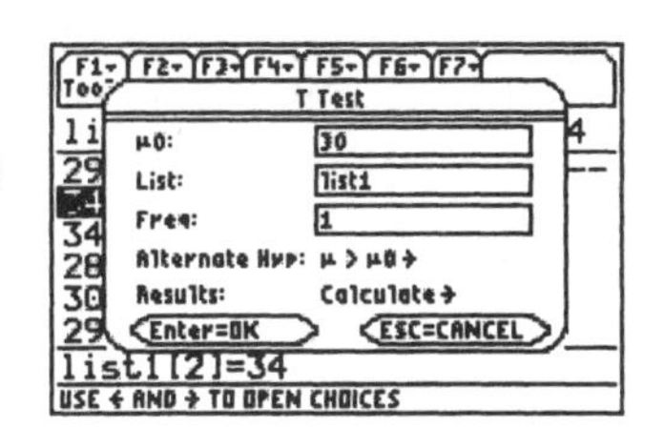

Selecting Draw yields the screen at right. We can clearly see the shaded portion of the curve which corresponds to the p-value for the test of 0.1257. The calculated test statistic is $t = 1.178$. The p-value indicates we'll expect to see a sample mean of 31.04 (the mean from our sample) or larger by chance about 12.5% of the time by randomness when the mean really is 30. That's not very rare. We fail to reject the null and conclude these data do not show motorists on the street are speeding, on average.

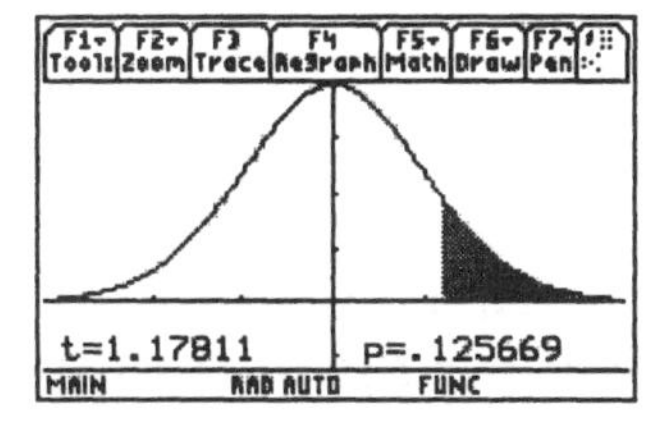

Comparing two means – confidence intervals

Should you buy name brand or generic batteries? Generics cost less, but if they do not last as long on average as the name brand spending the extra money may be worthwhile. Data were collected for six sets of batteries, which were used continuously in a CD player until no more music was heard through the headphones. The lifetimes (in minutes) for the six sets were:

Brand Name:	194.0	205.5	199.2	172.4	184.0	169.5
Generic:	190.7	203.5	203.5	206.5	222.5	209.4

The first step in performing a comparison such as this one (or any!) should always be to plot the data. Here, a side-by-side boxplot is natural.

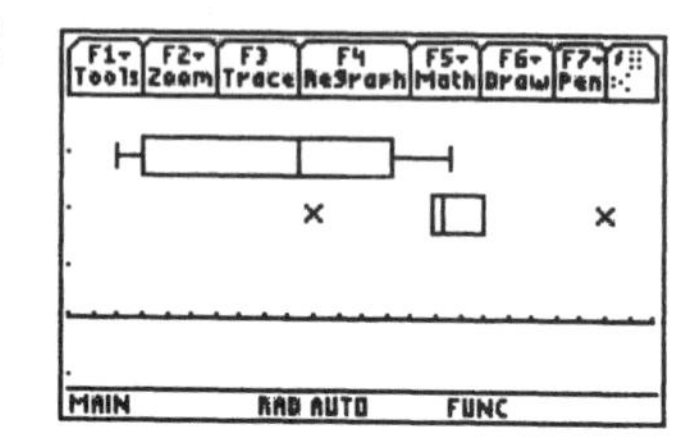

We have entered the data into `list1` for the Brand name batteries, and `list2` for the generics. We defined two boxplots to identify outliers on the `Plots` menu ([F2]). For more on these plots, see Chapter 16. From the plot, it certainly appears the generics last longer than the name brand batteries; they also seem more consistent (smaller spread). There are two outliers for the generic batteries, but with a sample size this small the outlier criteria are not very reliable. Neither of the extreme values are unreasonable, so it's safe to continue.

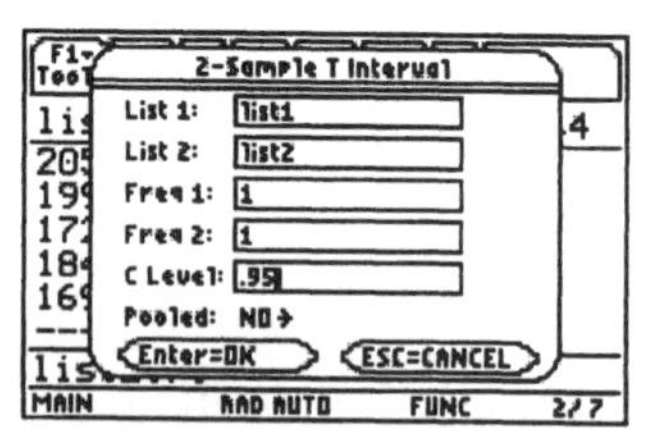

From the `Ints` menu select choice `4:2-SampTInt`. Our data are already entered, so move the highlight (if necessary) to `Data` and press [ENTER]. The data were in `list1` and `list2`, and each value in the lists occurred once. The confidence level has been set to 95% (entered as always in decimal form). The next option is new. `Pooled:` refers to whether the two groups are believed to have the same standard deviation or not. Visually, this is not true for our two battery samples. In general, unless there is some reason to believe the groups have the same spread, it's safest to answer this question with No. Reasoning behind this question has to do with computing a "pooled standard deviation" (or not) and the number of degrees of freedom for the test. Before the advent of computers (and statistical calculators) there were many recipes for handling this question, since the calculation of degrees of freedom in the unpooled case is complex. Luckily, we just let the calculator do the work.

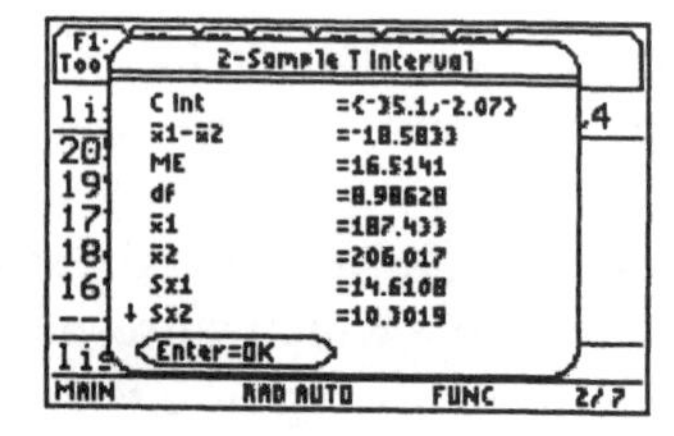

Pressing [ENTER] to calculate the interval gives the screen at right. We see we are 95% confident the average life of the name brand batteries is between 35.1 and 2.1 minutes *less* than the average life of the generic batteries. (Remember, it's always *group*1 – *group*2 in the interval). We are also given the difference in sample means and the margin of error. The next line gives the degrees of freedom for the interval – notice they're not even integer-valued. We also see the two sample means and standard deviations. The ↓ at the bottom left indicates more output can be obtained (the sample sizes). Assuming generic batteries are cheaper than name brand ones, it certainly would make sense to buy them.

Testing the difference between two means

If you bought a used camera in good condition, would you the same amount to a friend as to a stranger? A Cornell University researcher wanted to know how friendship affects simple sales such as this.[1] One group of subjects was asked to imagine buying from a friend whom they expected to see again. Another group was asked to imagine buying from a stranger. Here are the prices offered.

Friend:	\$275	300	260	300	255	275	290	300
Stranger:	\$260	250	175	130	200	225	240	

Here are side-by-side boxplots of the data. There certainly looks to be a difference. Prices to buy from strangers seem lower and much more variable than the prices for buying from a friend.

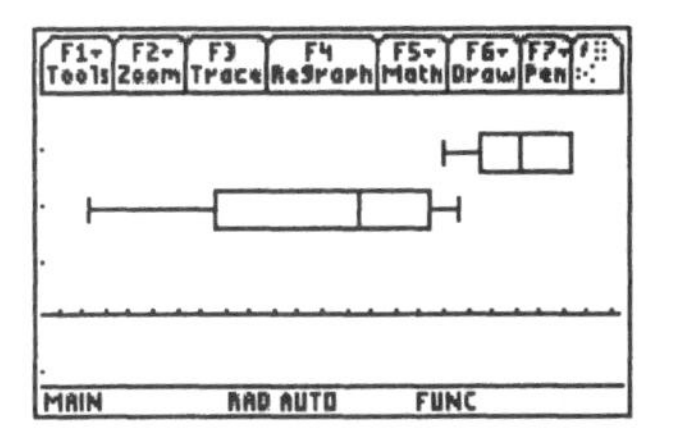

Returning to the Statistics List Editor, press [2nd][F1] to get the Tests menu and select choice 4 : 2-Samp Ttest. Again, we have the data in two lists, so Data is highlighted as the input mechanism, we have indicated the data are in list1 and list2, and each data value has a frequency of 1. The alternate hypotheses is $\mu_1 \neq \mu_2$ since our original question was "would you pay the same amount." Again, we have indicated No in regards to pooling the standard deviations (the spreads do not look equal and there is no reason to believe they should be the same). Notice the ⊙ in the lower left corner. Continuing to arrow down, we come to the choice of Calculate or Draw, which has been set to Calculate.

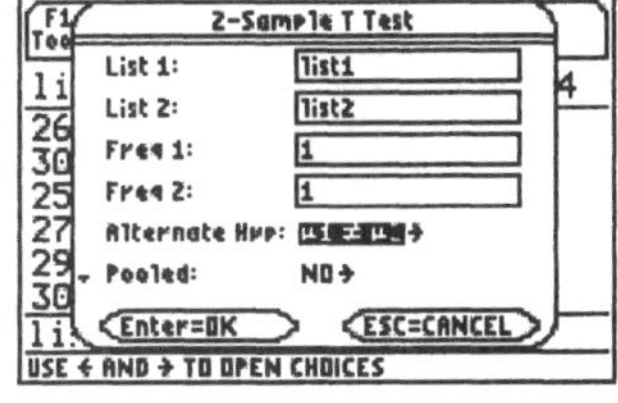

Pressing [ENTER] to calculate the test gives the screen at right. The computed test statistic is $t = 3.766$, and the p-value is 0.006. From these data we reject the null hypothesis of no difference and conclude that not only are people going to not pay the same amount to a friend than to a stranger, they're willing to pay more. We might even go so far as to warn people not to pay *too much* to friends.

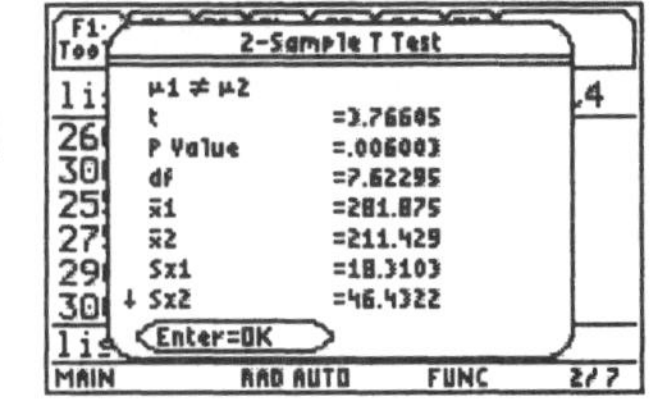

Paired Data

The two sample problems considered above used two *independent* samples. Many times data which might seem to be for two samples are naturally paired (say, examining the ages of married couples – each couple is a natural pair) or are even two observations on the same individuals. In such cases one works with the differences in each pair, and not the two sets of observations.

Do flexible schedules reduce the demand for resources? The Lake County (IL) Health Department experimented with a flexible four-day week. They recorded mileage driven by 11 field workers for a year

[1] Halpern, J.J. (1997). The transaction index: A method for standardizing comparisons of transaction characteristics across different contexts, *Group Decision and Negotiation*, 6(6), 557-572.

on an ordinary five-day week, then they recorded the mileage for a year on the four-day week.[2] Here are the data:

Name	5 day mileage	4 day mileage
Jeff	2798	2914
Betty	7724	6112
Roger	7505	6177
Tom	838	1102
Aimee	4592	3281
Greg	8107	4997
Larry G	1228	1695
Tad	8718	6606
Larry M	1097	1063
Leslie	8089	6392
Lee	3807	3362

Cursory examination reveals that after the change, some drove more, and some less. It is also easy to see there are large differences in the miles driven by the different workers. It is this variation between individuals that paired tests seek to eliminate.

We have entered the data into the calculator; the 5-day week mileages are in list1, and the 4-day mileages are in list2.

We need to find the differences. Move the cursor to highlight the name of list3. Using VAR-LINK, enter the command list1-list2 and press [ENTER]. The results will now be stored in list3.

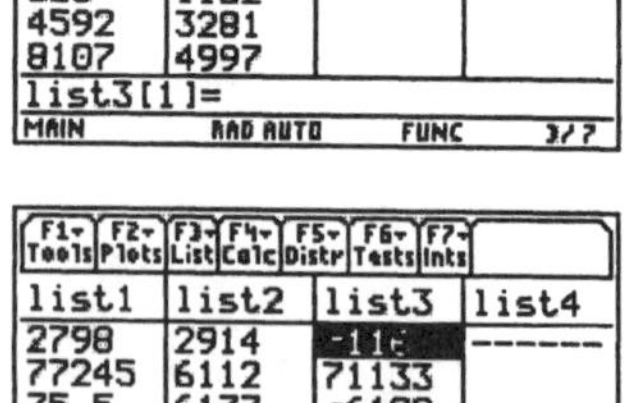

We need to check if the differences are approximately normal (or certainly at least no strong skewness or outliers). We define the normal plot as at right to use the differences we just created. Pressing [ENTER] will compute the *z*-scores. Return to the Plot Setup screen ([F2]) check that no other plots are turned on (if so, move the cursor to them and either clear the plot ([F3]) or uncheck them ([F4]). Press [F5] to display the graph.

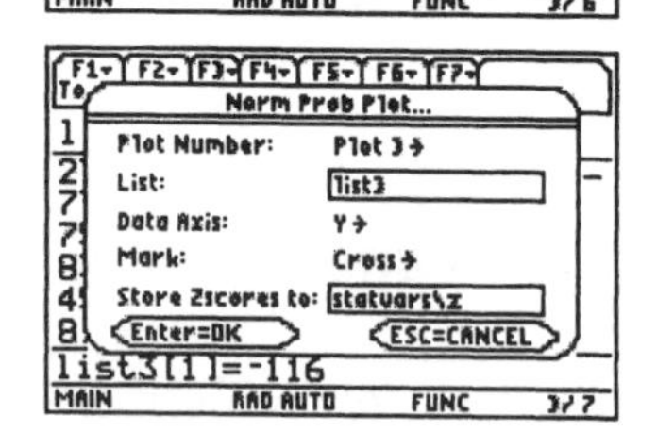

[2] Catlin, Charles S. Four-day Work Week Improves Environment, *Journal of Environmental Health*, Denver, March 1997 **59**:7.

The plot at right is not perfectly straight. However, there are no large gaps, so no extreme outliers.

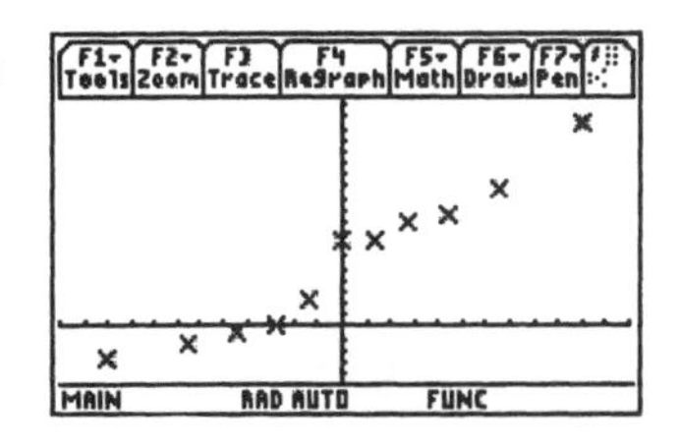

We now proceed to the test. We will perform a one-sample test using the differences as the data. From the Tests menu, select 2:T-Test. Make certain the Data Input Method is set to Data, and press [ENTER]. If the change in work week made no difference, the average value of the computed differences should be 0, so this is the value for μ_0. We are using the data from list3 as the input, and have selected the alternative hypothesis as $\mu \neq \mu_0$.

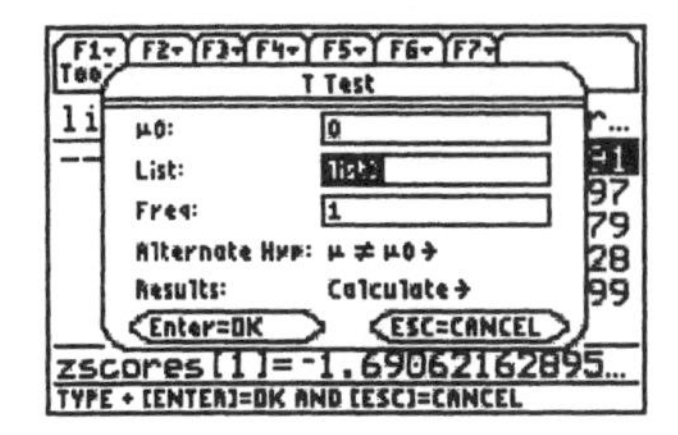

Pressing [ENTER] after selecting Calculate displays the results. The computed test statistic is $t = 2.85$ and the p-value is 0.017. We conclude that these data do indicate a difference in driving patterns between a 5-day work week and a 4-day work week. Further since the average difference is positive (982 miles) it seems that employees drove less on the 4-day week (the subtraction was 5-day – 4-day mileages). It's hard to say if the difference is meaningful to the department. If so, they may want to consider changing all employees to 4-day weeks.

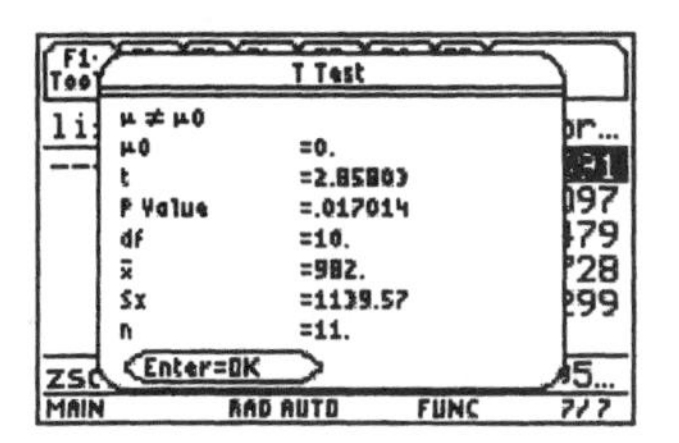

We can go further and compute a confidence interval for the average difference. Select 2:TInterval from the Ints menu, and define the interval as at right.

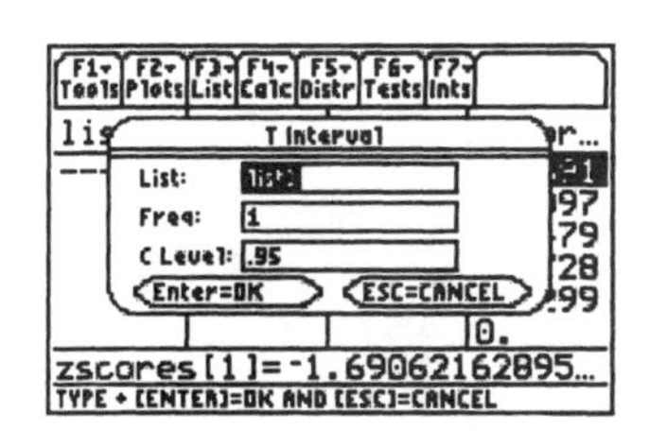

Pressing [ENTER] to calculate the interval, we find we are 95% confident the 5-day work week will average between 216.4 and 1748.3 more yearly miles than a 4-day work week.

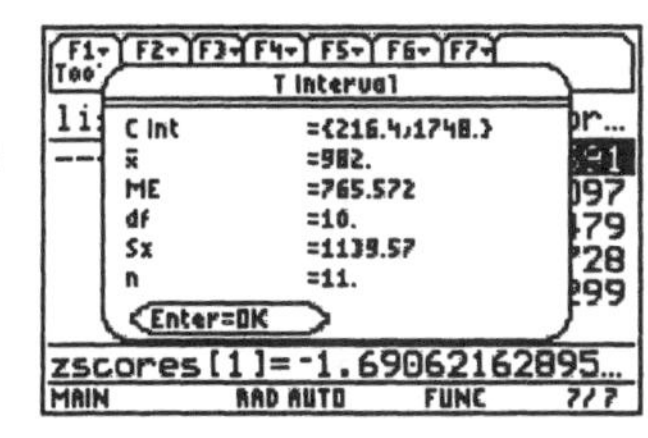

What can go wrong?

Not much that hasn't already been discussed – trying to subtract lists of differing length will give a dimension mismatch error. Having more plots "turned on" than are needed can also cause errors. The biggest thing to guard against is bad conclusions. Think about the data and what they show. Do not let conclusions contradict a decision to reject or not reject a null hypothesis.

Chapter 23 – Comparing Counts

Count data are analyzed primarily for three different purposes: whether or not data agree with a specified distribution (a goodness-of-fit test), whether or not observed distributions collected at different times or places are consistent with one another (a test of homogeneity), and whether or not data classified according to two categorical variables indicate the categorical variables are related or not (a test of independence). All of these tests use a probability distribution called the $\chi 2$ (chi-squared) distribution. The first test, that of goodness-of-fit, is built into the TI-89 as are the other two tests for which the mechanics are exactly the same. What is different is the setting and conclusions which can be made.

The $\chi 2$ statistic is defined to be $\sum \frac{(Obs - Exp)^2}{Exp}$ where the sum is taken over all the cells in a table. The quantities $\frac{(Obs - Exp)}{\sqrt{Exp}}$ are the standardized residuals which should be examined in the event the null hypothesis is rejected to determine which cells deviated most from what is expected.

Testing Goodness-of-fit

Does your zodiac sign determine how successful you will be in later life? Fortune magazine collected the zodiac signs of 256 heads of the largest 400 companies. Here are the number of births for each sign.

Births	Sign
23	Aries
20	Taurus
18	Gemini
23	Cancer
20	Leo
19	Virgo
18	Libra
21	Scorpio
19	Saggitarius
22	Capricorn
24	Aquarius
29	Pisces

We can see some variation in the number of births per sign, but is it enough to claim that successful people are more likely to be born under certain star signs than others? If there is no difference between the signs, each should have (roughly) 1/12 of the births. That's the null hypothesis in this situation: births are evenly distributed across the year. The alternate is that the null is wrong: births are not evenly distributed across the year.

Here, the observed counts have been entered into `list1` and the expected counts in `list2`. The expected count for each sign in this problem is 256/12. The calculator displays the fraction in reduced form. In general, the expected cell count will be n (the grand total) times the probability for each cell. Alternatively, expected counts can be entered in decimal form.

list1	list2	list3	list4
	64/3		
20	64/3		
18	64/3		
23	64/3		
20	64/3		
19	64/3		

list1[1]=23

Note: make sure both lists are the same length!

Press [2nd][F1](F6) for the `Tests` menu. Select choice `7:Chi2 GOF`. You will enter the list name containing the observed counts, the list name containing the expected counts, and the degrees of freedom which are k-1 where k is the number of categories. You have the choice (as always) of simply computing the test or drawing the distribution curve to display the p-value graphically.

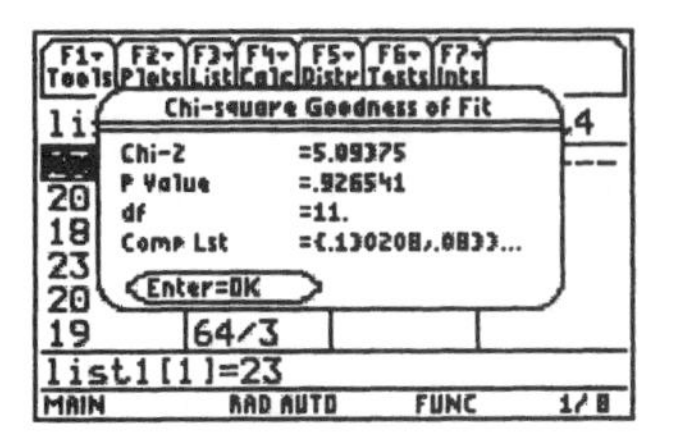

Having selected Calculate and pressing [ENTER] displays the results. The computed value of the $\chi 2$ statistic is 5.094 and the p-value for the test is 0.9265. With this large p-value the null hypothesis is not rejected, and we conclude the distribution of signs is evenly spread throughout the year.

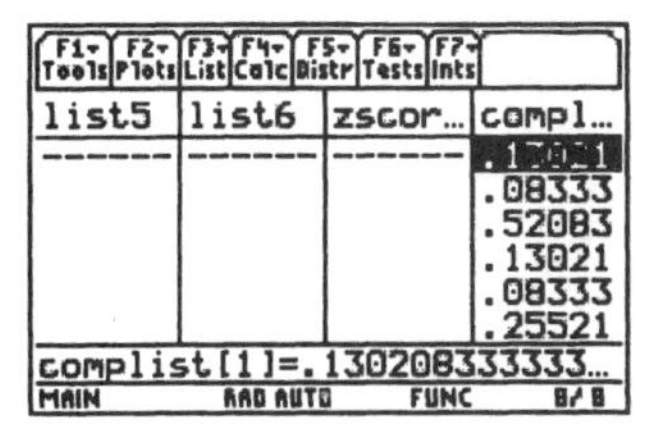

The calculator automatically stores a new list of the components of the $\chi 2$ statistic. The square roots of these values are the standardized residuals (after appropriate positive and negative signs are attached). Scanning down this list we see (not surprisingly) the birth sign most different from its expected value is Pisces, the last entry in the list.

Alternatively, the standardized residuals can be computed explicitly. If the cursor highlight is placed in the list name (here, list3) entering the command $(list1 - list2)/\sqrt{list2}$ will place these values into the new list.

Tests of Homogeneity

Many high schools survey graduating classes to determine their plans for the future. We might wonder whether plans have stayed roughly the same or have changed through time. Here is a summary table from one high school. Each cell of the table shows how any student from each graduating class (the columns) had that particular type of plan (the rows). (Source IHS)

	1980	1990	2000
College/Post HS Educ	320	245	288
Employment	98	24	17
Military	18	19	5
Travel	17	2	5

Visually, choices do not appear to be the same (look at the row for Employment) but is the difference real or is it due perhaps to different size classes? Since we have really the same distribution at different time points, this is a test of homogeneity: the null hypothesis is that the distribution of students' plans is the same across time, the alternate hypothesis is that the distributions are not the same.

We will first enter the numbers in the body of the table into a matrix. From the home screen, press [APPS] Arrow to choice `6:Data/Matrix Editor` and press the right arrow. Select choice `3:New`.

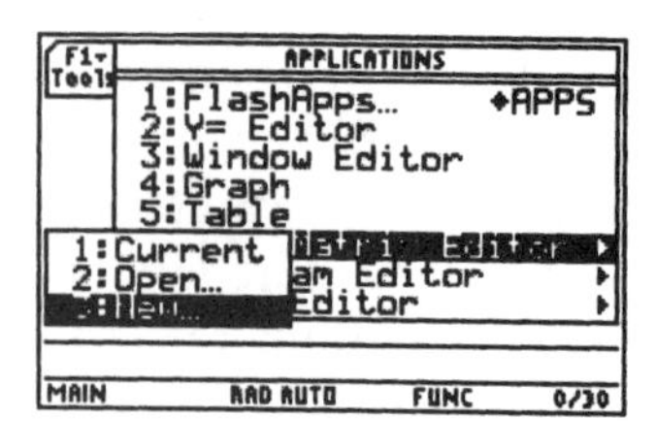

We first define the type. Press the right arrow and select `Matrix`. Leave the folder set as main unless you really want to change it. Press the down arrow and give the matrix a name. The cursor in this field is set to alpha by default, so pressing any number keys will result in their letter equivalent. If a number is desired, press [alpha] to change out of alpha mode. The body of our table had 4 rows and 3 columns, so the matrix is 4 x 3. Press [ENTER].

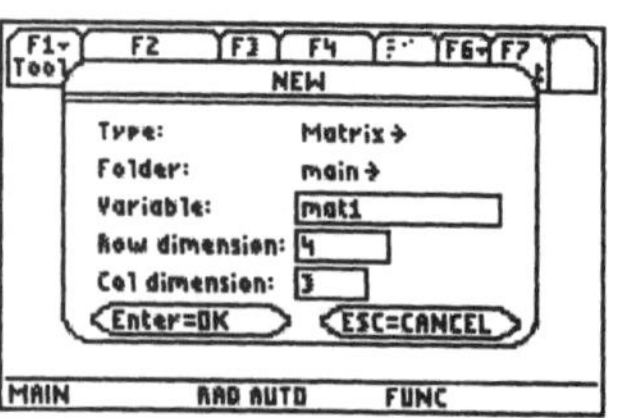

Cells are entered across rows. Press [ENTER] after each cell entry to predeed to the next. Here is the filled-in matrix. Press [2nd][ESC] (QUIT) to leave the Matrix Editor.

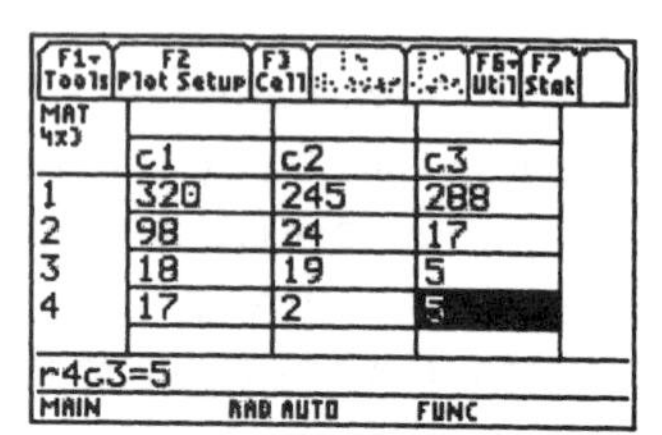

Return to the Statistics app and press [2nd][F1] for the `Tests` menu. Select choice `8:Chi2 2-way`. To enter the name of the observed matrix, Go to the `VAR-LINK` screen ([2nd][-]) and find the name you gave it. The calculator will automatically store expected counts and components of the $\chi 2$ statistic in the named statistics variables.

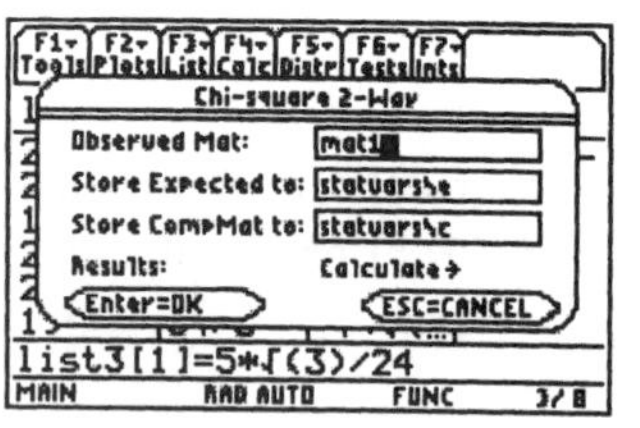

Selecting `Draw` gives the screen at right. We see three things here: the $\chi 2$ statistic (72.7738) and the p-value (1.1×10^{-13}). With so small a p-value no area appears shaded. We also see a different sort of distribution curve. $\chi 2$ curves are not symmetric; they are right skewed.

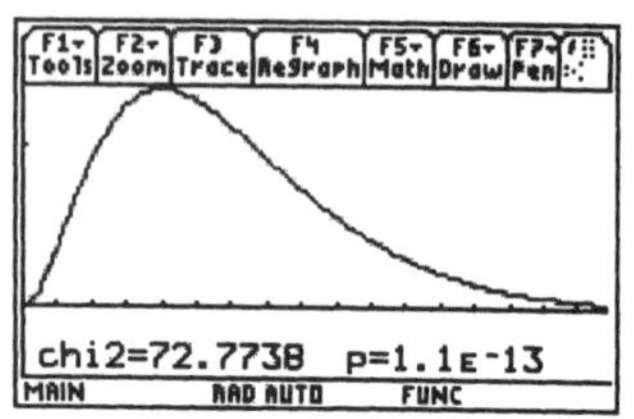

If we had chosen `Calculate` instead of `Draw` we would have this screen. It shows the same statistic value, but a little more exact p-value. We are also given the degrees of freedom for the test which are $(rows - 1)(cols - 1)$, so (4-1)(3-1) = 6. We also see the first components in the Expected Counts and Components matrices.

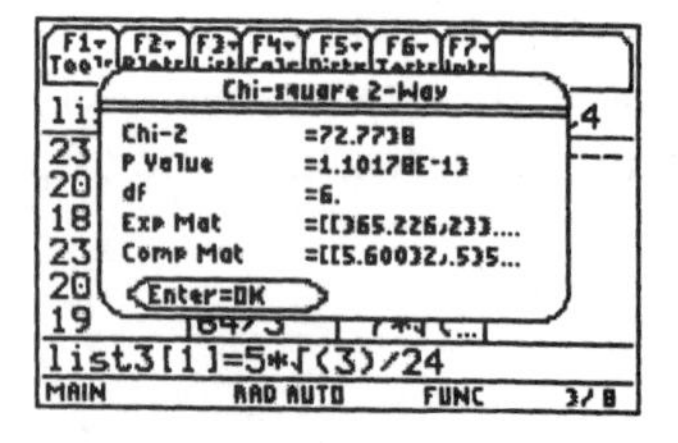

Since the p-value is so small, we reject the null hypothesis and conclude the distributions of post-graduation plans are not the same for all three years.

Where are the differences? We'd like to get the standardized residuals. Unfortunately, the TI-89 won't give these easily. We can look at the components of the $\chi 2$ statistic. The standardized residuals are the (signed) square roots of these values. Large entries in this matrix do indicate which cells differ greatly from what is expected under the null hypothesis of independence.

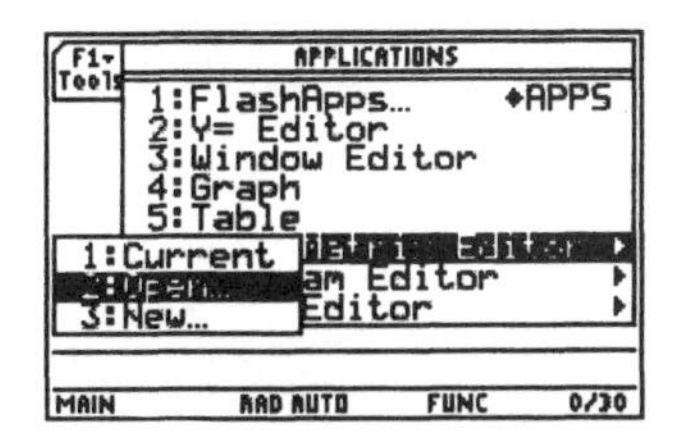

From the home screen, press [APPS]. Select 6:Data/Matrix Editor and then 2:Open. Press [ENTER].

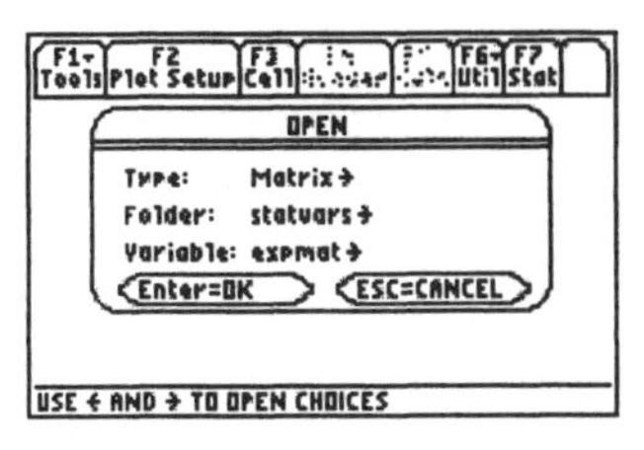

The first thing to do might be to check the matrix of expected counts and compare them to the observed counts.The type we want is Matrix. The folder is statvars. The variable is expmat. Press the right arrow on each entry to get a list of choices. Use the down arrow to move between the main options. Press [ENTER] when finished to display the matrix.

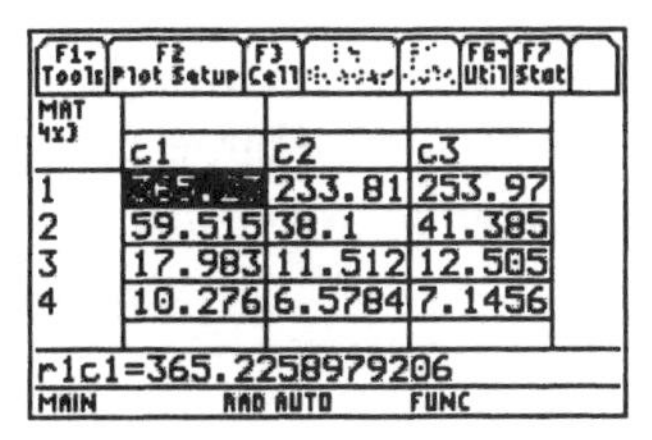

This is the matrix of expected counts. We can compare these to the observed counts in the original table. We can clearly see the actual count for 1980 who planned on Employment (98) is much larger than expected (58.5). The actual number in 2000 who planned on Employment (17) is much less than expected (41.4).

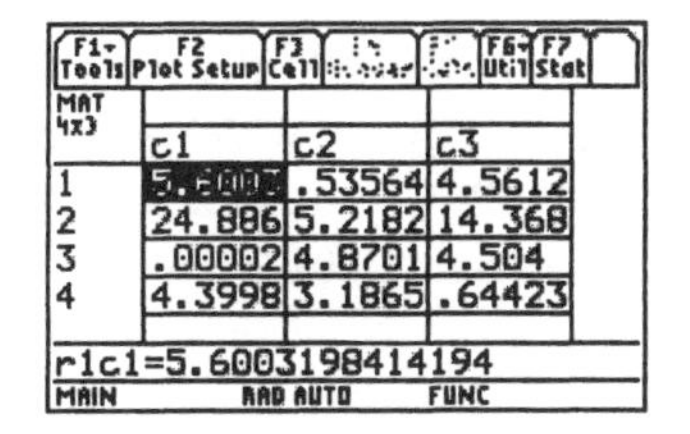

In a similar manner we can get the matrix of components, compmat.
The largest entries are for graduates whose plans were for employment in 1980 and 2000. The standardized residual for the 1980 entry is $\sqrt{24.886} = 4.989$ which we know is positive from the observed value being larger than the expected. Similarly, the standardized residual for the 2000 Employment entry is –3.791. There seems to be a decreasing trend in the number of graduates planning on employment.

Testing Independence

Tests of independence are used when the same individuals are classified according to two categorical variables. A study from the University of Texas Southwestern Medical Center examined whether the risk of Hepatitis C was affected by whether people had tattoos and by where they got their tattoos. The data from this study can be summarized in a two-way table as follows.

	Hepatitis C	**No Hepatitis C**
Tattoo, Parlor	17	35
Tattoo, Elsewhere	8	53
No Tattoo	22	491

Is the chance of having hepatitis C independent of (not related to) tattoo status? Our null hypothesis is that the two are not related. The alternate hypothesis is that there is a relationship.

Enter the numbers from the body of the table into a 3 x 2 matrix as discussed above. Press [2nd][ESC] (Quit) to exit the matrix editor.

The $\chi2$ test is performed just as it is for the test of homogeneity. Here are the results. The p-value is extremely small. We will reject the null hypothesis and conclude there is a relationship between hepatitis C status and tattoos.

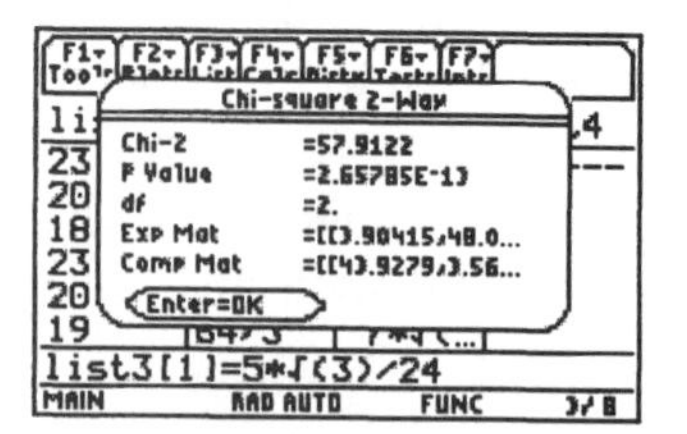

This is the matrix of components. People with tattoos from tattoo parlors make by far the largest contribution to the $\chi2$ statistic. Perhaps tattoo parlors are a source of hepatitis C?

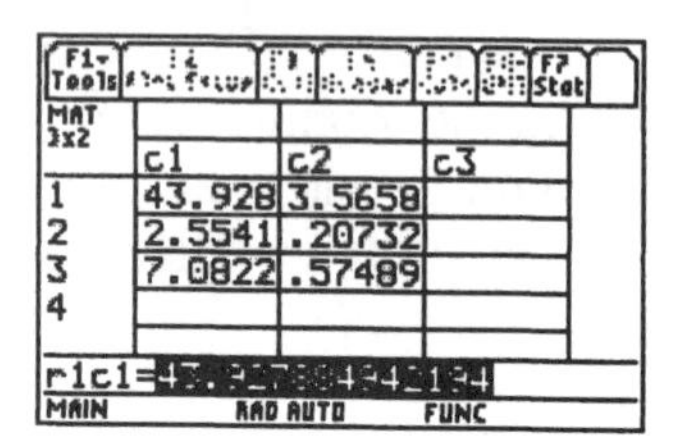

Here is the matrix of expected cell counts. Not all of them are more than 5. This means the conclusions based on the $\chi2$ test we just performed may not be valid. Since the largest standardized residual is for one of these cells, this is a problem. A common solution is to combine cells in some manner to overcome the problem. In this case, both rows for tattooed people could be combined into one.

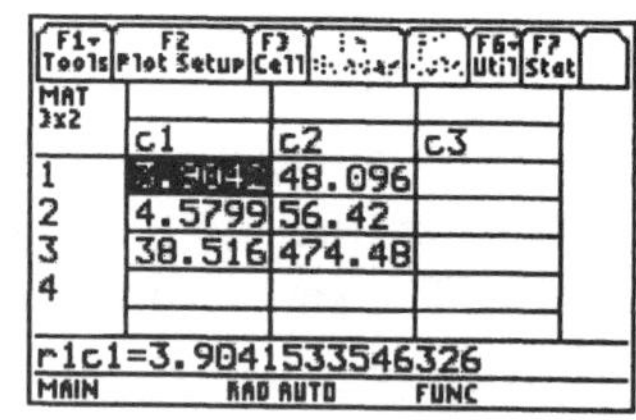

What can go wrong?

Expected cell counts less than 5.

Check the computed matrix of expected cell counts. If they are not all greater than 5 the analysis may be invalid.

Overusing the test.

These tests are so easy to do and data from surveys and such are commonly analyzed this way. The problem that arises here is that in this situation the temptation is to check many questions to see if relationships exist; but performing many tests on *dependent* data (the answers came from the same individuals) such as this is dangerous. In addition, remember that, just by random sampling, when dealing at $\alpha = 5\%$ we'll expect to see something "significant" 5% of the time when it really isn't. This danger is magnified when using repeated tests - it's called the problem of multiple comparisons.

Chapter 24 – Inference for Regression

Computing a regression equation and looking at residuals plots is not the end of the story. We might want to know if the slope (or correlation) is meaningfully different from 0. It's not always apparent that a slope is meaningfully non-zero. Consider these two equations for the selling price of a house: $price = 25 + 0.061 * sqft$ and $price = 25000 + 61 * sqft$. At first blush one might look at the small value for the slope in the first and believe it's reasonable to say the true slope may in fact be 0; however the difference is in the units – the first has price measured in thousands of dollars, the second in dollars. They're really the same line. In addition, we'd like to (perhaps) make a confidence interval for a "true" slope just as we did for means and proportions as well as confidence intervals for the average value of y for a given x and prediction intervals for a new y observation for an x value.

Returning to a problem considered before, here are advertised horsepower ratings and expected gas mileage for several 2001 vehicles.

Audi A4	170 hp	22 mpg	Buick LeSabre	205	20
Chevy Blazer	190	15	Chevy Prism	125	31
Ford Excursion	310	10	GMC Yukon	285	13
Honda Civic	127	29	Hyundai Elantra	140	25
Lexus 300	215	21	Lincoln LS	210	23
Mazda MPV	170	18	Olds Alero	140	23
Toyota Camry	194	21	VW Beetle	115	29

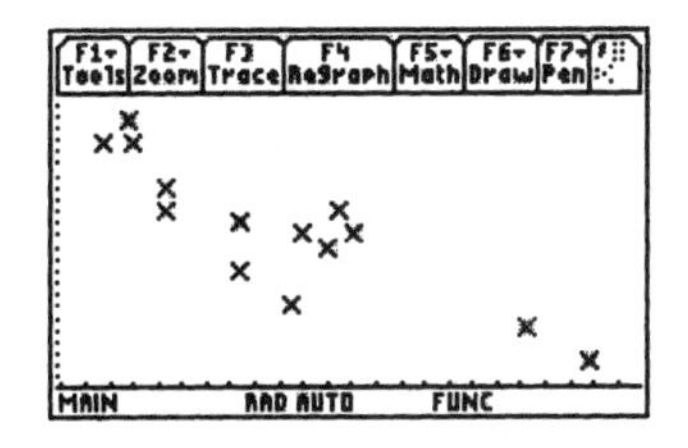

How is horsepower related to gas mileage? Recall the plot that was constructed for this data in chapter 5. It is reproduced at right. The trend is decreasing. The residuals plots in Chapter 5 showed no overt pattern against X (horsepower) and the normal probability plot was reasonably straight. Inference for the regression is therefore appropriate.

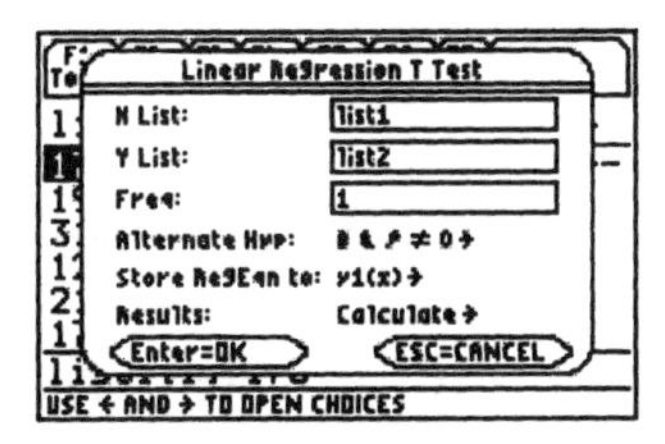

From the Statistics list editor, press [2nd][F1] (F6, `Tests`) and select choice `A:LinRegTTest`. You tell the calculator which list contains the x (predictor variable) values, which contains the y (response) values. `Freq` is normally set to 1. Indicate the appropriate form of the alternate hypothesis. Notice there is an option to store the equation of the line. To store the equation as a function press the right arrow key and select a function name. Finally, there is an option to simply calculate the t-test or to draw the t distribution curve and shade the area corresponding to the p-value of the test.

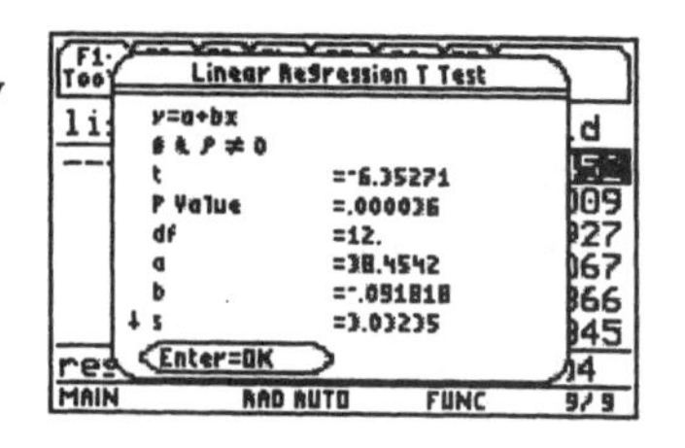

This is the first portion of the output (notice the ↓ at the bottom left). The first lines indicate the form of the regression so that you are reminded which quantity is the slope (b) and which the intercept (a) and the form of the alternate hypothesis in the test. The computed t-statistic for this regression is –6.35 and the p-value for the test is 0.00004, with 12 degrees of freedom. We will reject the null hypothesis and conclude not only that the slope is not zero; it is significantly negative. The intercept for the regression is 38.45. . The slope is -0.092 which we know is significantly different from zero even though its value seems small. The standard deviation of the data points around the line is 3.03.

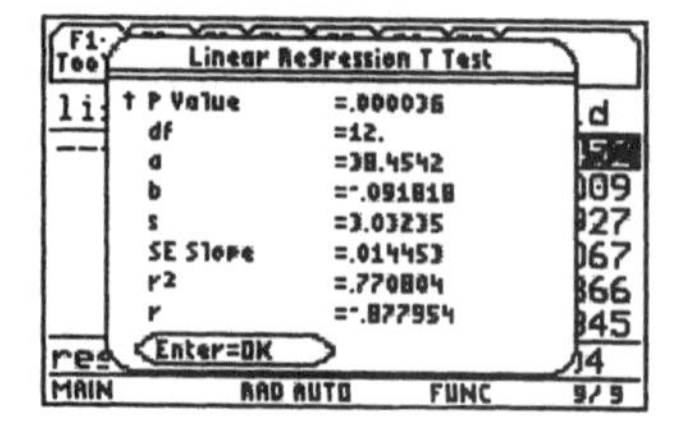

Pressing the down arrow several times we find the rest of the output. The standard error of the slope ($s(b_1)$) is 0.0145. The relationship is strongly negative since r^2 = 77.1% and r = -0.877

A confidence interval for the slope

Confidence intervals (for any quantity) are always $estimate \pm (criticalvalue)(SE(estimate))$. In this case the critical value of interest will be a t statistic based on 12 degrees of freedom. From tables, we find this is 2.179 for 95% confidence. The standard error of the slope is $SE(b_1) = \dfrac{s(e)}{\sqrt{\sum (x-\bar{x})^2}} = \dfrac{s(e)}{\sqrt{n-1} * s(x)}$, which was already found in the regression output to be 0.145, rounded to three places.
Putting all the pieces together, the 95% confidence interval for the slope is $-0.092 \pm 2.179 * 0.0145$ or (-0.124, -0.060). Based on this regression, we are 95% confident average gas mileage decreases between –0.124 and –0.060 miles per gallon for each horsepower in the engine.

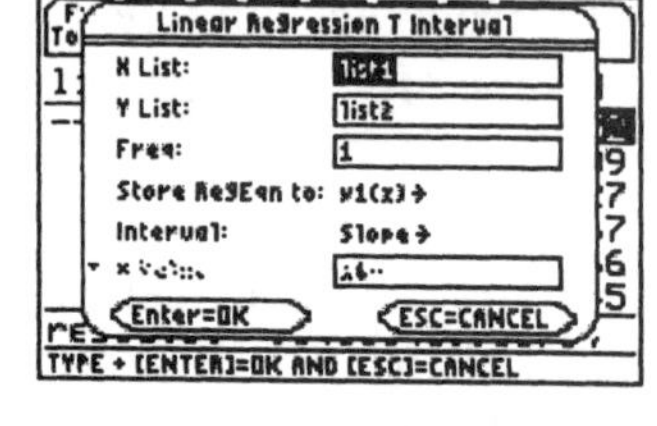

Alternatively, the TI-89 will calculate this interval for you. From the Statistics list editor, press [2nd][F2] (F7, Ints). Select menu choice 7:LinRegTInt. You have a choice on this menu after specifying where the lists of data are to form intervals for the slope or response. Here, we want an interval for the slope.

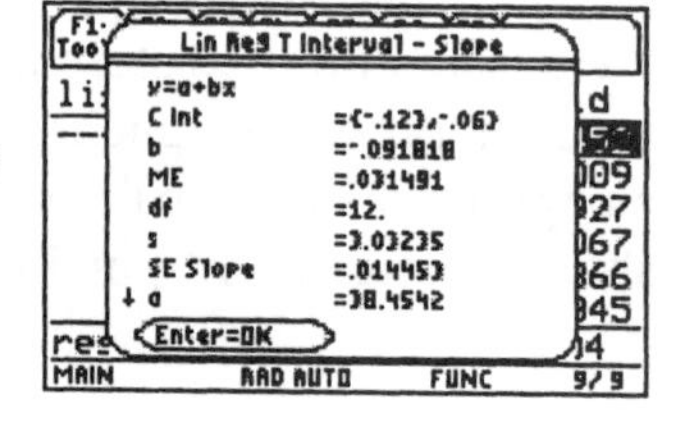

The results indicate the interval is –0.123 to –0.06. The slight difference between the calculator's values and those obtained above is due to rounding errors. The calculator is generally more accurate as it keeps all significant digits in performing calculations whereas people tend to round in intermediate steps. Pressing the down arrow on this screen will give the values of r and r^2.

A confidence interval for the mean at some X

What should we predict as the average gas mileage for a vehicle with 160 horsepower? Evaluating the equation of the line for 160 horsepower gives 23.76 miles per gallon. This is just a point estimate, however and is subject to uncertainty just as any mean is. Confidence intervals account for this uncertainty – in this case there are two sources – average variation around the line as well as uncertainty about the slope which makes estimation more "fuzzy" further away from the mean. Both of these are accounted for in the equation of the standard error, $SE(\hat{\mu}_v) = \sqrt{s^2(b_1) * (x_v - \bar{x})^2 + \frac{s^2(e)}{n}}$. Putting everything together, we find

$SE(\hat{\mu}_v) = \sqrt{0.0145^2 * (160 - 185.429)^2 + \frac{3.0323^2}{14}} = 0.890$. The t critical value is still 2.179, so the confidence interval is 23.76 ± 2.179*0.890 or (21.82, 25.70). Based on this regression, we estimate with 95% confidence the average gas mileage for vehicles with 160 horsepower will be between 21.82 and 25.70 miles per gallon.

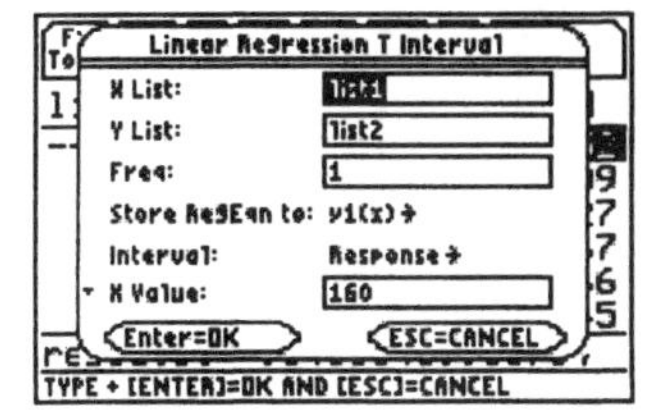

The calculator can also find this interval. From the Statistics list editor, press [2nd][F2] (F7, Ints). Select menu choice 7:LinRegTInt. On the input screen select Response for the interval type and enter the X value desired. Press [ENTER] to perform the calculation.

We are first shown y_hat (the value from the line for 160 horsepower) which is 23.76, and the degrees of freedom (12). The confidence interval for the mean response is given as 21.82 to 25.7. The margin of error and standard error are also given.

A prediction interval for a new observation

What would we predict for gas mileage for a particular vehicle with 160 horsepower? The point estimate is still 23.76 miles per gallon, but we have some additional uncertainly because individual observations are more variable than means. The standard error becomes $SE(\hat{y}_v) = \sqrt{s^2(b_1) * (x_v - \bar{x})^2 + \frac{s^2(e)}{n} + s^2(e)}$ which becomes $SE(\hat{\mu}_v) = \sqrt{0.0145^2 * (160 - 185.429)^2 + \frac{3.0323^2}{14} + 3.0323^2} = 3.159$. So the prediction interval is 23.76 ± 2.179*3.159 or (16.88, 30.64). Based on this regression we estimate with 95% confidence the gas mileage for a vehicle with 160 horsepower will be between 16.88 and 30.64 miles per gallon.

Obtaining this interval from the calculator is found as described above for the mean response interval; simply press the down arrow to find this interval immediately after the confidence interval.

What can go wrong?

Assuming lists are the same length, not much that has not already been covered. One problem in doing many of these computations "by hand" comes from the compounding of round-off errors in intermediate computations. One is generally safest in using many digits in the interim and rounding only at the end. (Notice the "hand calculated" intervals are somewhat different from those obtained from the calculator. This is the reason

Chapter 25 – Analysis of Variance (ANOVA)

We have already seen two-sample tests for equality of the means in Chapter 22. What if there are more than two groups? Answering the question relies on comparing variation among the groups to variation within the groups, hence the name. The null hypothesis for ANOVA is always that all groups have the same mean and the alternate is that at least one group has a mean different from the others.

Wild irises are beautiful flowers found throughout North America and northern Europe. Sir R. A. Fisher collected data on the sepal lengths in centimeters from random samples of three species. The data are below. Do these data indicate the mean sepal lengths are similar or different?

Iris setosa	Iris versicolor	Iris virginica
5.4	5.5	6.3
4.9	6.5	5.8
5.0	6.3	4.9
5.4	4.9	7.2
5.8	6.7	6.4
5.7	5.5	5.7
4.4	6.1	
	5.2	

I have entered the data into lists list1, list2, and list3. We will first construct side-by-side boxplots of the data for visual comparison. Visually, the medians are somewhat different with Iris virginica being the largest.

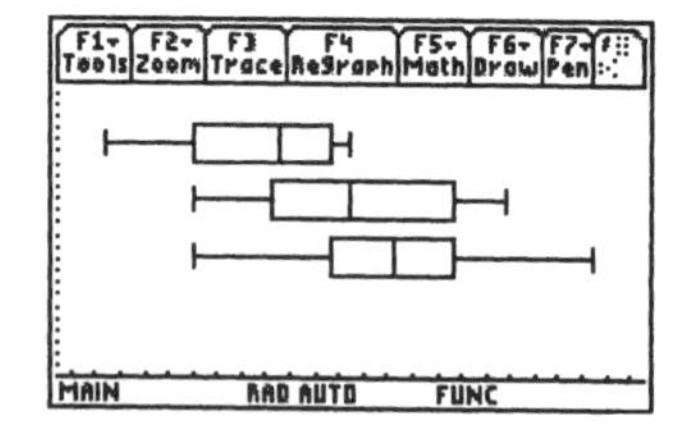

From the STAT Tests menu select choice C:ANOVA. This is the next-to-last test on the menu, so it is easiest to find by pressing the up arrow.

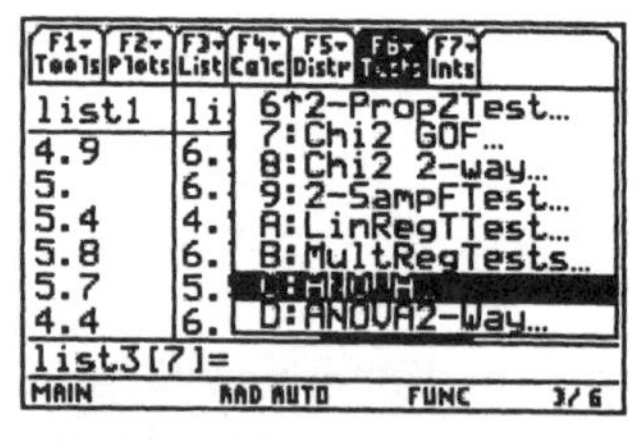

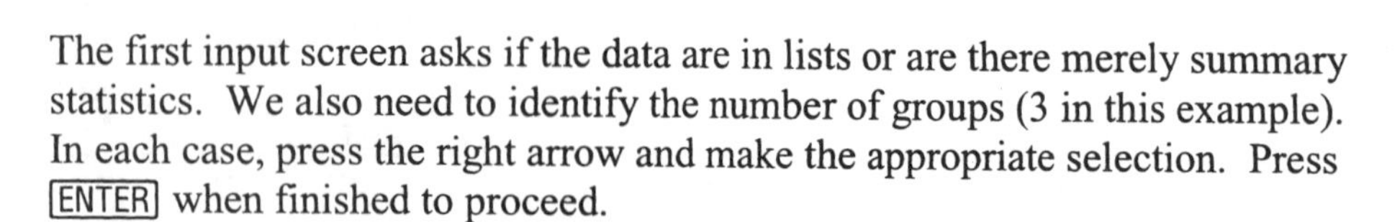
The first input screen asks if the data are in lists or are there merely summary statistics. We also need to identify the number of groups (3 in this example). In each case, press the right arrow and make the appropriate selection. Press ENTER when finished to proceed.

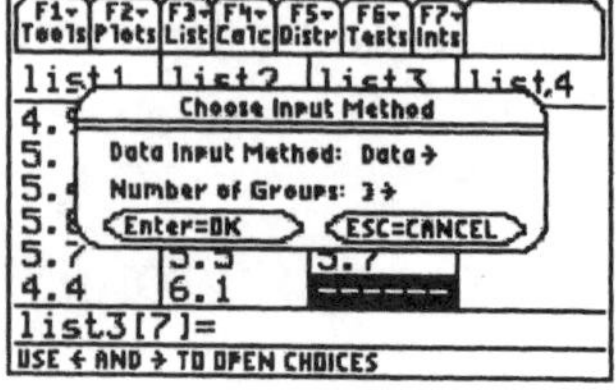

The second input screen asks for the list names. Use [2nd][–] (VAR-LINK) to access the list of list names and select those you have used.

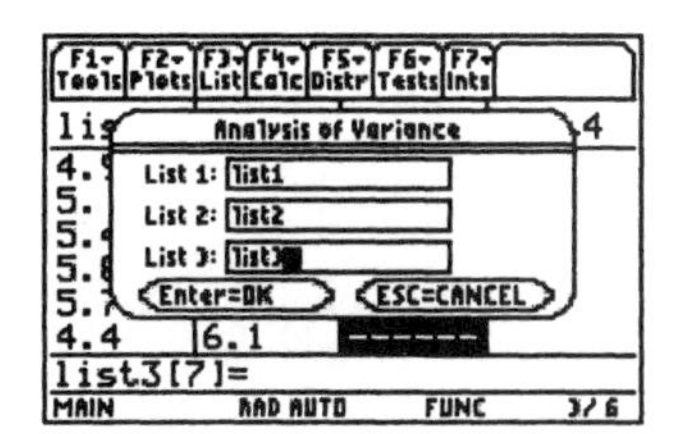

This is the first portion of the output. The value of the F statistic is 2.95 and the p-value for the test is 0.0779 which indicates at $\alpha = 5\%$ there is not a significant difference in the mean sepal lengths for the three species, based on this sample. The Factor degrees of freedom are k-1 where k is the number of groups, so with three groups, this is 2. MS is SS/df.

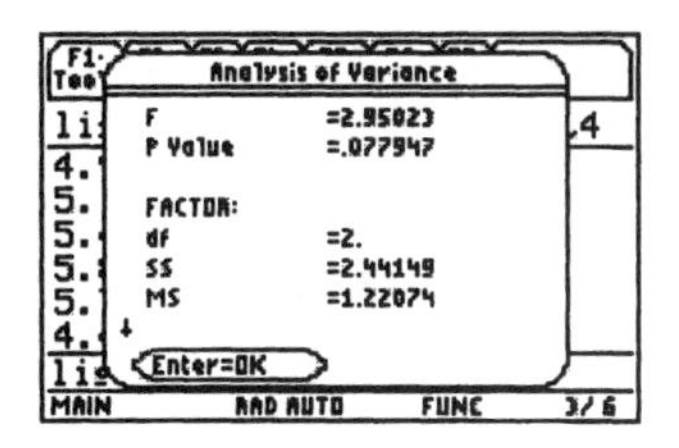

Pressing the down arrow several times gives the remainder of the output. Degrees of freedom for Error are n-k, where n is the total number of observations in all groups (21 here) and k is the number of groups (3). MS is again SS/df. Sxp is the estimate of the common standard deviation and is the square root of MSE. The F statistic is MSFactor/MSError.

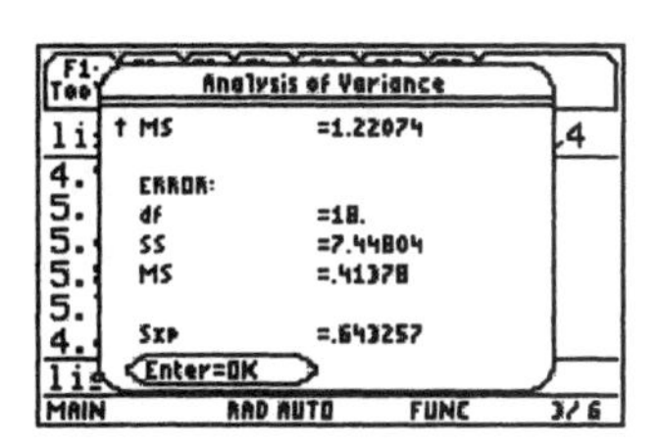

After pressing [ENTER] to clear the output screen, we find something has been added into the Statistics editor. The first column contains the means for each group. The next two lists, lowlist and uplist, contain the lower and upper limits of individual 95% t-intervals for the mean, where Sxp has been used as the common standard deviation. Looking at these is a crude method of determining whether group means are equal or not. If the intervals have a large amount of overlap it is an indication that group means are not different. Since all of these intervals overlap, it is a confirmation of the decision that the mean sepal lengths of the three species are similar.

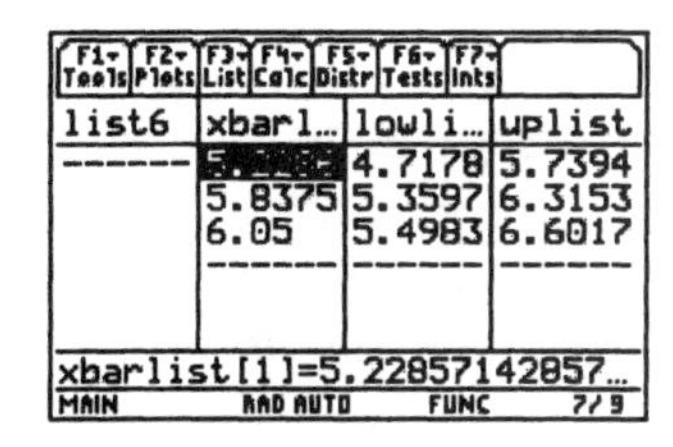

Another Example

The following data represent yield (in bushels) for plots of a given size under three different fertilizer treatments. Does it appear the type of fertilizer makes a difference in mean yield?

Type A	Type B	Type C
21	41	35
24	44	37
31	38	33
42	37	46
38	42	42
31	48	38
36	39	37
34	32	30

Here are side-by-side boxplots of the data. All three distributions appear symmetric and there is not a large difference in spread, so ANOVA is appropriate. Using `1-Var Stats` we find the mean for Type A is 32.125 bushels, Type B has a mean of 40.125 bushels, and Type C has a mean of 37.25. They're different, but are they different enough?

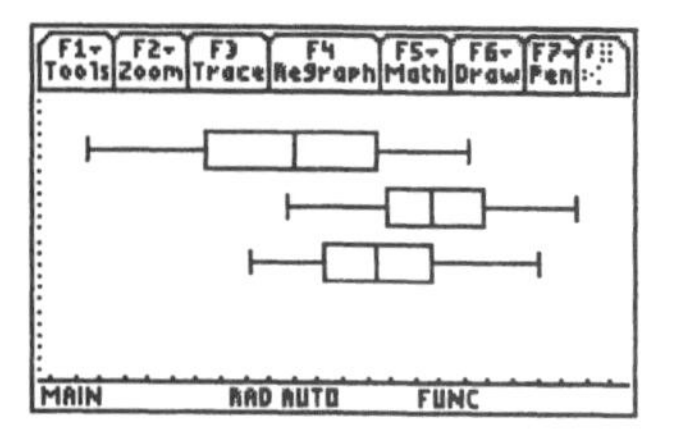

The ANOVA output of interest is at right. We see the F-statistic is 4.05 and the p-value for the test is 0.0325. At $\alpha = 0.05$, we will reject the null hypothesis and conclude that at least one fertilizer has a different mean yield than the others. Scrolling down, we further find Sxp = 5.696.

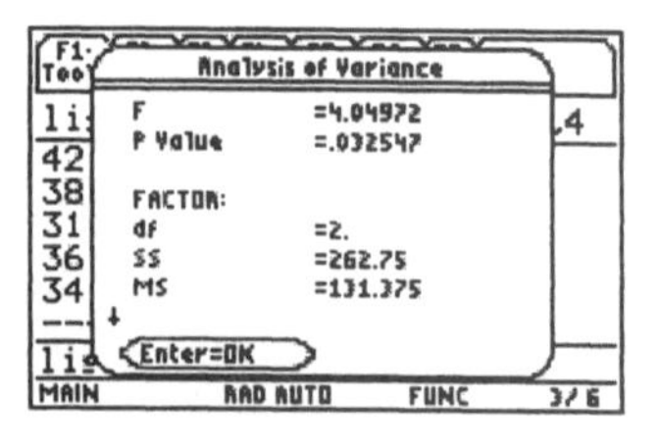

Here we see why the individual confidence intervals are a crude measure of which means are different. Our test concluded at least one fertilizer had a different mean from the others. All of the intervals overlap! (However, the intervals for types A and B don't overlap by much – maybe they're the ones that are different?)

list6	xbar1...	lowli...	uplist
------	32.125	27.937	[highlighted]
	40.125	35.937	44.313
	37.25	33.062	41.438

uplist[1]=36.312755476198

MAIN RAD AUTO FUNC 9/9

Which mean(s) is (are) different?

Having rejected the null, we would like to know which mean (or means) are different from the rest. We've already seen that individual confidence intervals don't always tell the true story. Similarly, testing each pair of means (doing three tests here) has the same problem: the problem of multiple comparisons. If we constructed three individual confidence intervals, the probability they all contain the true value is $0.95^3 = 0.857$, using the fact that the samples are independent of each other.

The Tukey method is a way to solve this question. From tables, we find the critical value q^* for k groups and n-k degrees of freedom for error. My table gives $q^*_{3,21,0.05} = 3.58$ The "honestly significant difference" is computed as $HSD = \frac{q^*}{\sqrt{2}} s_p \sqrt{\frac{2}{n_i}}$ where each group has n_i observations. (NOTE: most tables of this distribution require the division of q^*, but not all. Check your particular table.) Here, $HSD = \frac{3.58}{\sqrt{2}} * 5.695 \sqrt{\frac{2}{8}} = 7.208$. Groups with means that differ by more than 7.208 are significantly different. The mean for Type A was 32.125; for Type B 40.125 and for Type C 37.25. We therefore declare types A and C have similar means; Types B and C also have similar means. The means which are different from each other are those for types A and B.

ANOVA with summary statistics

The TI-89 is also capable of doing the analysis with summary statistics instead of data, just like testing one or two means.

A student decided to investigate just how effective washing with soap is in eliminating bacteria. To do this she tested four different methods – washing with water only, washing with regular soap, washing with antibacterial soap and spraying hands with antibacterial spray (active ingredient 65% Ethanol). She suspected that the amount of bacteria on her hands before washing might vary considerably from day to day. To help even out the effects of those changes, she generated random numbers to determine the order of the treatments. Each morning she washed her hands according to the treatment randomly chosen and then placed her right hand on a sterile media plate. The plate is designed to encourage bacterial growth. She incubated each plate for 2 days at 36°C after which she counted the bacterial colonies. She replicated each treatment 8 times. Her data are below.

Level	Number	Mean	Std Dev
Alcohol Spray	8	37.5	26.56
Antibacterial Soap	8	92.5	41.96
Soap	8	106.0	46.96
Water	8	117.0	31.13

We will test the null hypotheses that type of soap makes no difference in the average number of bacteria against the alternate that at least one method is better than the others (fewer bacteria, on average).

Here, we have selected the data input method is Stats, and told the calculator there are four groups.

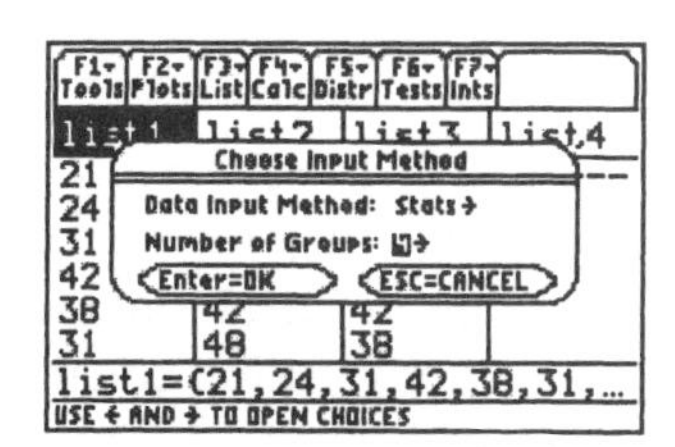

The statistics (n, $\bar{x}$, and s) are input for each group enclosed in curly braces ([2nd][(] and [2nd][)]) and separated by commas. Don't be alarmed if some of the entry disappears to the left of the input box.

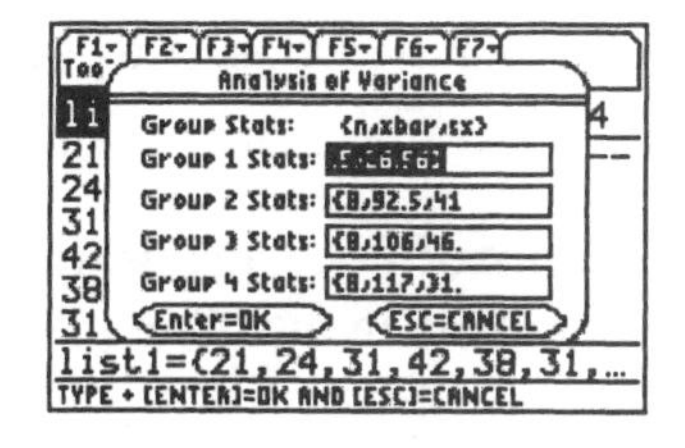

Here is the first portion of the output. With an F statistic of 7.06 and a p-value of 0.0011, we reject the null hypothesis and conclude the different soap types leave different amounts of bacteria, on average. The calculator does not give confidence intervals for the individual means when using this input method.

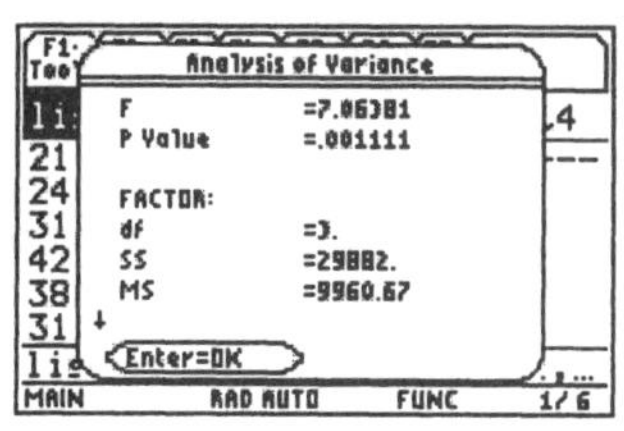

What can go wrong?

Syntax.

This error is caused by failure to input the statistics enclosed in curly braces. Pressing [ESC] will return you to the data input screen so the problem can be corrected.

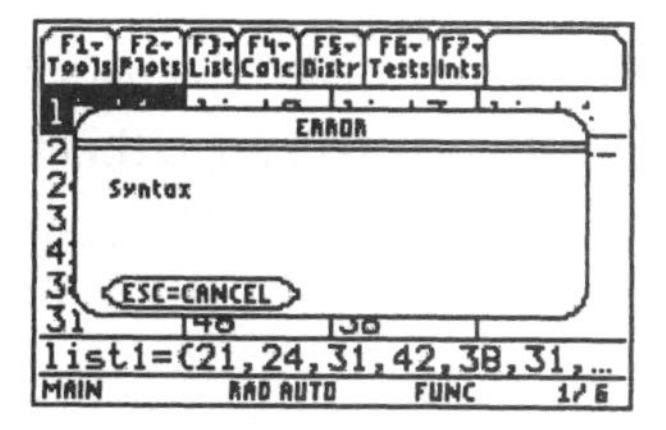

Chapter 26 – Multiple Regression

Multiple regression is an extension of the linear regression already studied where we create a model to explain a response variable based on more than one predictor. Just as with linear regression, we will want to examine how well the predictors individually and as a group determine the response by testing the utility of the model and create confidence intervals for slopes, mean response, and predictions of new responses.

How well do age and mileage determine the value of a used Corvette? The author chose a random sample of ten used Corvettes advertised on autos.msn.com. The data are below.

Age (Years)	Miles (1000s)	Price ($1000)
3	46	27
1	11	43
2	20	35.5
1	11.5	39
8	69	16.5
5	49	23
2	10	38
4	27	32
5	30.5	30
3	46	27

We first examine plots of each predictor variable against Price. The plot against age is linear, and decreasing as expected (we expect older cars to cost less). The regression equation for this relationship is $Carprice = 42.44 - 3.33 * Age$, with $r^2 = 80.6\%$. This suggests the average price of a used Corvette goes down $3330 each year.

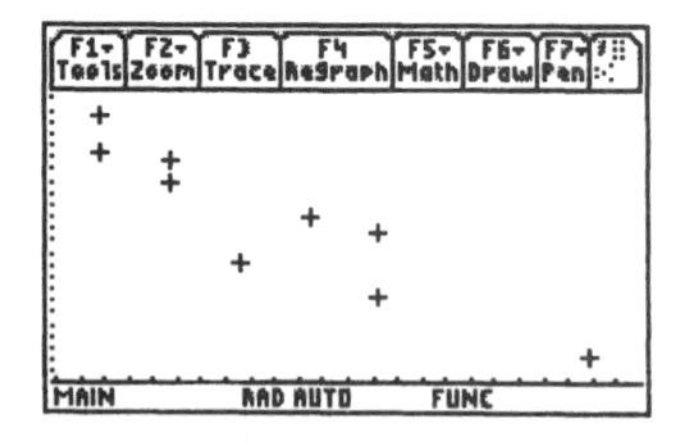

The plot of Price against mileage is also linear and decreasing with even less scatter than in the other plot. The regression equation for this relationship was found to be $Carprice = 43.77 - 0.40 * Miles$, with $r^2 = 95.5\%$. This suggests the average price of a used Corvette will decrease $400 for every 1000 miles it has been driven.

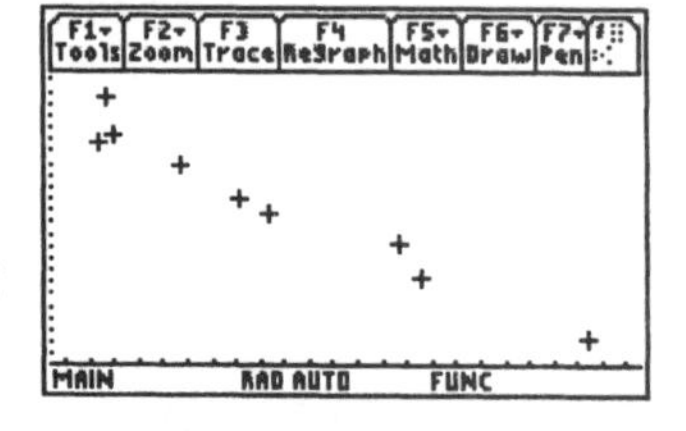

From the Statistics list editor, press [2nd][F1] (Tests). Select menu option B:MultRegTests.

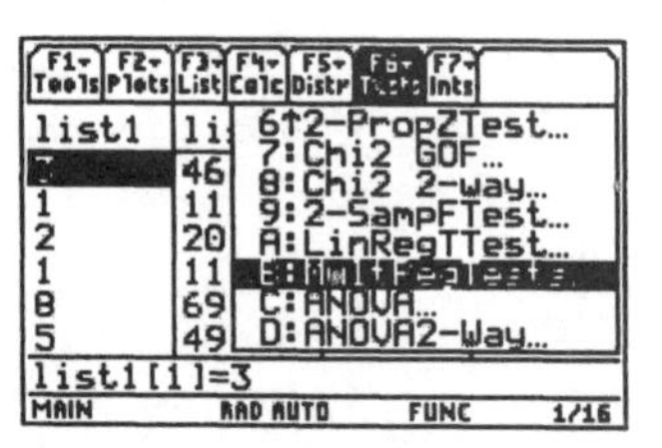

The input screen first asks how many independent variables there are. Use the right arrow to change this to the proper number. Enter the list names for the Y list and the X lists using [2nd][-] (VAR-LINK). Press [ENTER] to perform the calculations when all list names have been entered.

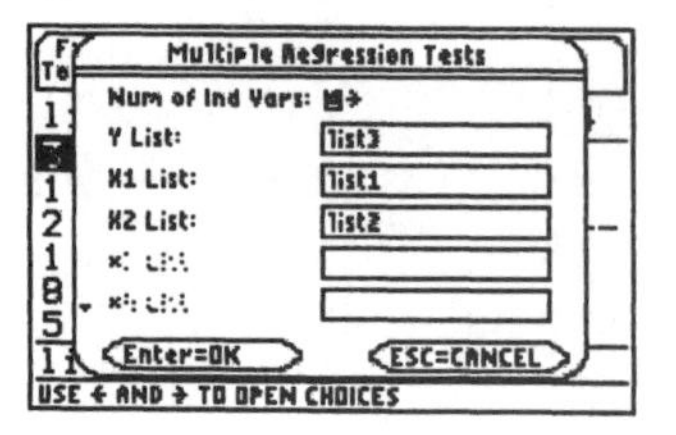

Here is the first portion of the output. The fist line is the form of the regression equation. The second and third lines give the F statistic for the overall significance (utility) of the model, along with its p-value. Here the p-value of 0.000003 indicates there is a significant relationship between price and its two predictors. R^2 in this situation is the amount of variation in price which is explained by the two predictors (97.3%).

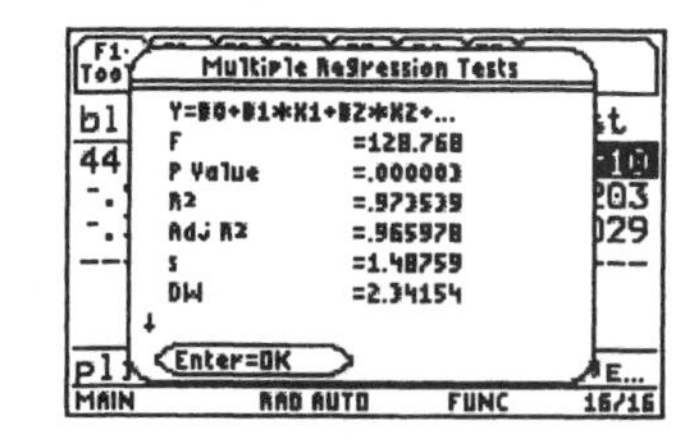

The next quantity is the adjusted R^2. Since R^2 and never decrease when additional variables are added into a model, this quantity is "penalized" for additional variables which do not significantly help explain variation in Y, so it will go down if this is the case. Adjusted R^2 is always less than the regular R^2 . *S* is the standard deviation of the residuals.

DW is the value of the Durbin-Watson statistic which measures the amount of correlation in the residuals and is useful for data which are time series (data has been collected through time). If the residuals are uncorrelated, this statistics will be about 2 (as it is here); if there is strong positive correlation in the residuals, DW will be close to 0; if the correlation is strongly negative, DW will be close to 4. Since these data are not a time series, DW is meaningless for our example.

Pressing the down arrow we find the components for regression and error which are used in computing the F statistic. The F statistic for regression is the MS(Reg)/MS(Error) where the Regression Mean square functions just like the treatment (factor) mean square in ANOVA.

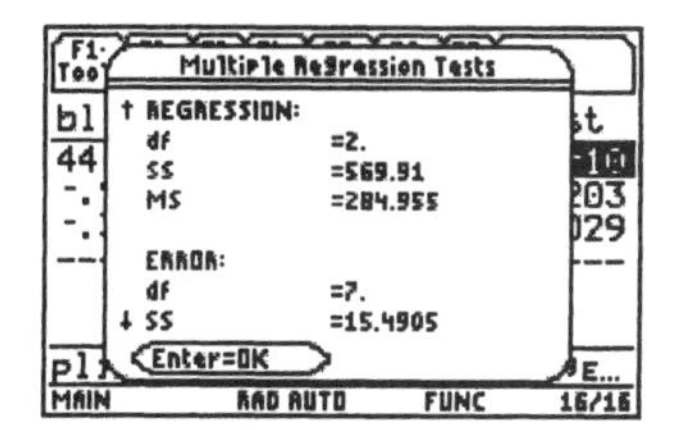

Finally we see some of the entries in new lists that have been created. The complete lists will be seen when [ENTER] is pressed. `Blist` contains the estimated intercept and coefficients; `SE list` is the list of standard errors for the coefficients which can be used to create confidence intervals for true slopes;

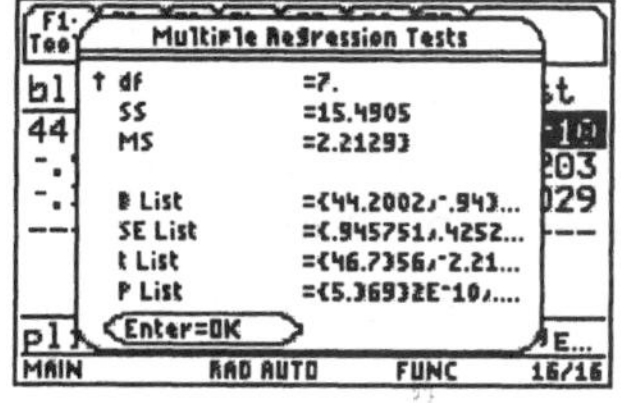

`t list` gives values of the *t*-statistics for hypothesis tests about the slopes and intercept; `P list` gives the p-values for the tests of the hypotheses H_0: $\beta_i = 0$ against H_A: $\beta_i \neq 0$. If the assumed alternate is 1-tailed, divide these p-values by 2 to get the appropriate p-value for your test.

After pressing [ENTER] we see several new lists that have been added into the editor. `Yhatlist` is the list of predicted values for each observation in the dataset based on the model ($yhat_i = \beta_0 + \beta_1 x_{1i} + \beta_2 x_{2i}$ in this model); `resid` is the list of residuals $e_i = y_i - yhat_i$. `Sresid` is a list of standardized residuals obtained by dividing each one by *S*, since they have mean 0. If the normal model assumption for the residuals is valid, these will be N(0, 1).

zscor...	yhatl...	resid	sresid
-------	[illegible]	-.1494	-.1283
	39.856	3.1438	2.4267
	36.131	-.6305	-.4591
	39.702	-.7016	-.5409
	15.322	1.178	1.2537
	24.335	-1.335	-.9918

yhatlist[1]=27.1493826141...

Pressing the right arrow we find yet more lists. Leverage is a measure of how influential the data point is. These values range from 0 to 1. The closer to one, the more influential (more of an outlier in its *x* values) the point is. Values greater than 2p/n where *n* is the number of data points and *p* is the number of parameters in the model are considered highly influential. Here,

lever...	cookd	blist	selist
[illegible]	.00348	44.2	.94575
.24161	.62537	-.9434	.42529
.14756	.01216	-.3091	.04637
.23948	.0307	------	------
.601	.78915		
.18114	.07253		

leverage[1]=.387715535021...

$n = 10$ and $p = 3$, so any value grater than 0.6 will designate an observation as highly influential. This indicates in our example that the point for the eight-year-old car is influential.

Cook's Distance in the next column is another measure of the influence of a data point in terms of both its x and y values. Its value depends on both the size of the residual and the leverage. The ith case can be influential if it has a large residual and only moderate leverage, or has a large leverage value and a moderate residual, or both large residual and leverage.

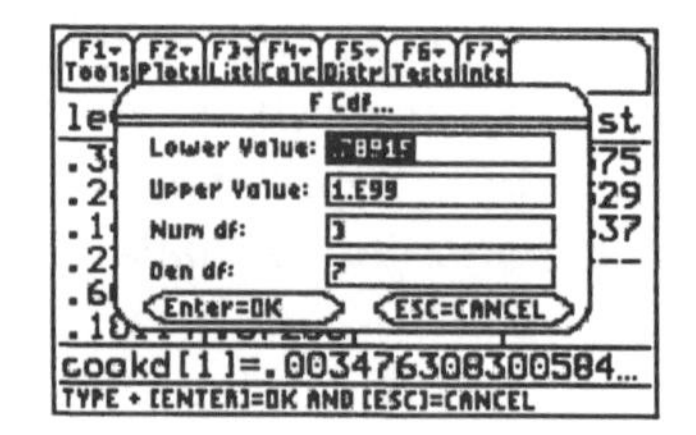

To assess the relative magnitude of these values, one can compare them against critical values of an F distribution with p and $n - p$ degrees of freedom or use menu selection A: F Cdf from the [F5] (Distr) menu. The largest value in the list is for the Corvette data is again for the 5th observation (the 8-year-old). This is the input screen. We are finding the area above 0.78915 which is the largest value in the list. The result is 0.5372, which indicates this is not unusual, so this point is not influential.

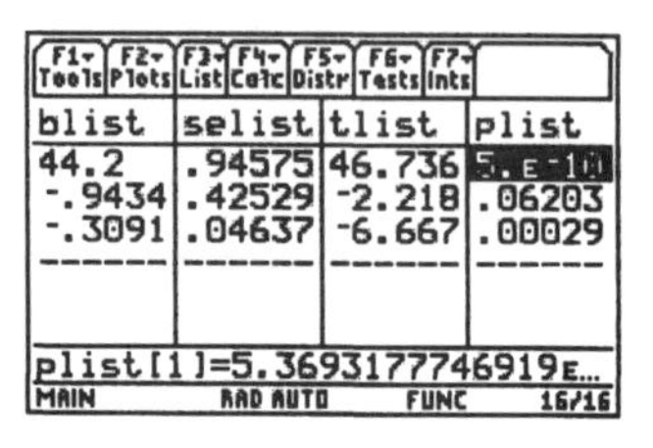

After pressing the right arrow still more, we find the last of the output lists. Blist is the list of coefficients. We finally see the fitted regression equation: $Price = 44.2 - 0.94 * Age - 0.31 * Miles$. We interpret the coefficients in the following manner: Price declines \$940 on average for each year of age when mileage is the same; for cars of a given age, every additional 1000 miles reduces average price \$310. The coefficient of miles is similar to that obtained from the simple regression (\$400) but the decrease for age is much less than the value for the simple regression (\$3300).

The next column contains the standard errors of each coefficient. These can be used to create confidence intervals for the true values using critical values for the t distribution for $n - p$ degrees of freedom. Finally we see the t statistics and p-values for testing H_0: $\beta_i = 0$ against H_A: $\beta_i \neq 0$. These suggest the coefficient of age is not significantly different from 0; in other words, mileage is a much more determining quantity for the price of used Corvettes which helps explain why its coefficient changed less than the coefficient of age from the single variable regressions.

Assessing the model

Just as with simple (one-variable) linear regression, we will use residuals plots to assess the model. We plot the residuals (either standardized or not) against the fitted values (not the individual x variables) and check a normal plot.

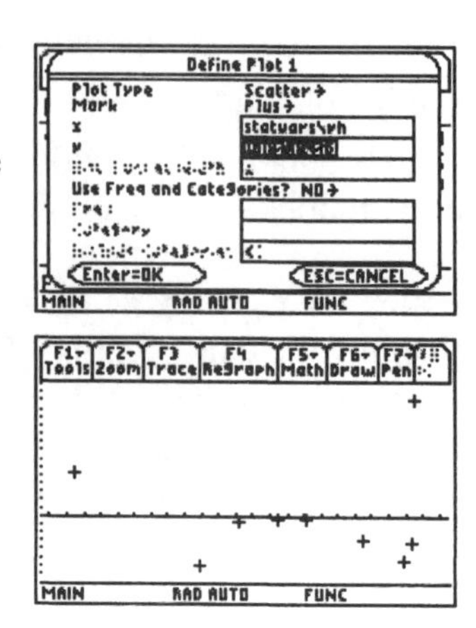

We first define a scatter plot of residuals against fitted values. The list names are accessed from the VAR-LINK menu ([2nd][-]). To find them, arrow down until you reach the beginning of the STATVARS portion, then press Y to move to that portion of the list, find and select yhatlist for the X variable, repeat the process pressing R ([2]) to move to the r portion of the list and find and select resid (or S ([3])and select sresid) as the Y variable. The plot definition is at right.

Pressing [ENTER] followed by [F5] displays the plot. Just as with simple regression we are looking in this plot for indications of curves or thickening/narrowing which indicate problems with the model. With this small data set these are somewhat hard to see, but clearly the only positive residuals are for the largest and smallest fitted values, which could indicate a potential problem.

We define a normal plot of the residuals (as in chapter 4). Before displaying this plot, be sure to either clear the scatter plot or uncheck it using the Plot Setup screen (F2 from the Statistics List editor). This normal plot is not a straight line, which indicates a violation of the assumptions. This multiple regression model is not appropriate for these data.

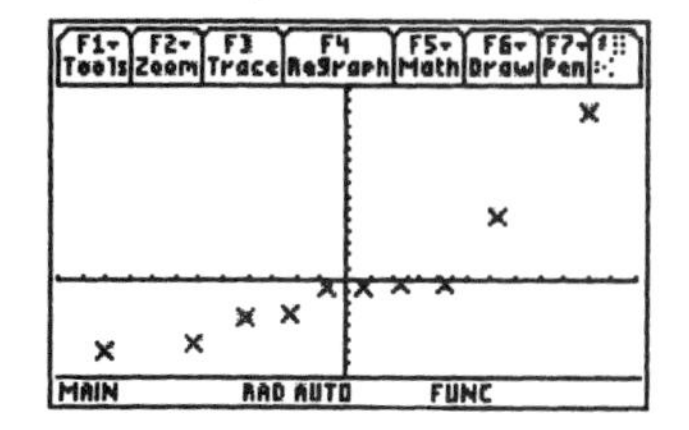

Multiple Regression Confidence Intervals

How well do the midterm grade and number of missed classes predict final grades? Data for a sample of students are below.

Final Grade, Y	Midterm Exam, x_1	Classes Missed, x_2
81	74	1
90	80	0
86	91	2
76	80	3
51	62	6
75	90	4
48	60	7
81	82	2
94	88	0
93	96	1

Performing the regression as above, we find the estimated regression equation $Final = 49.41 + 0.502 * Midterm - 4.71 * ClassesMissed$. All coefficients are significantly different from zero and a normal plot of the residuals is relatively straight. We'd like to use the model to create a prediction of the average grade for students with a midterm grade of 75 and 2 absences (a confidence interval); we also want to predict the grade for a particular student with a midterm grade of 75 and 2 absences (a prediction interval).

The data are entered with Final grades in `list1`; midterm grades in `list2`, and absences in `list3`. Order of specification in the regression input screen matters. In `list4` we have entered the values for the predictions of interest. They must be entered in the order in which the xlists will be specified.

Press 2nd F2 (F7) and select option `8:MultRegInt`. We are first asked the number of independent variables. In our data we have 2. Use the right arrow to access the list of possible values, and select the appropriate one for your data set. Press ENTER to continue.

Now specify the lists to be used in the regression and the list containing the values to be used in the intervals as at right. Specify the desired confidence level, here 95%, as a decimal.

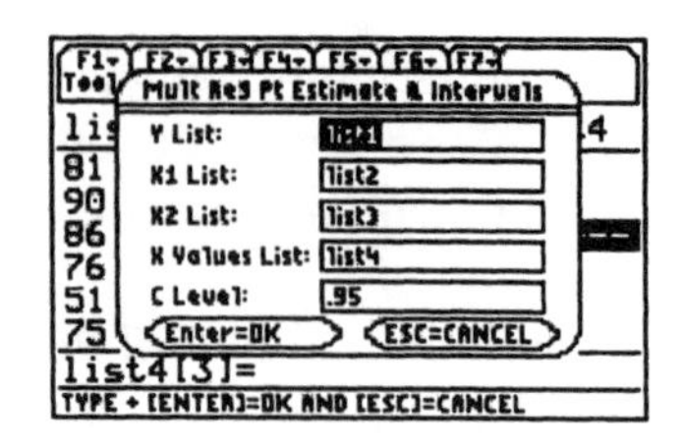

Pressing [ENTER] performs the calculations and displays the screen at right. We find the point estimate of the final grade for a student with a 75 midterm grade and two absences is 77.66 or 78 (practically speaking). The confidence interval says we are 95% confident the average final grade for all students with a 75 midterm grade and 2 absences will be between 76.7 and 78.6.

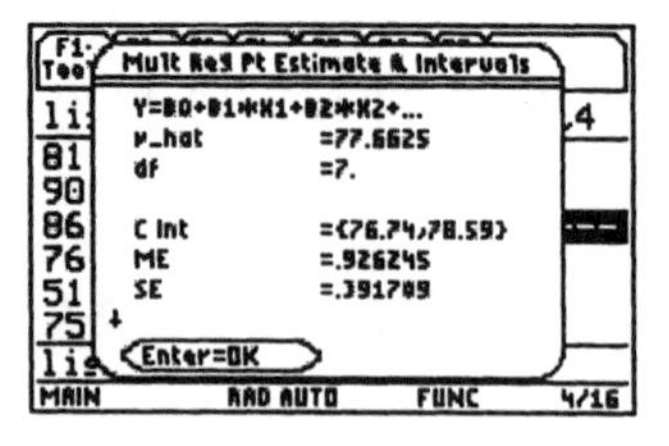

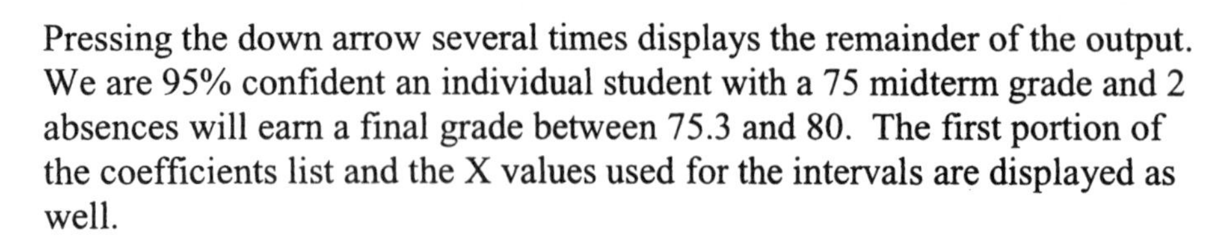
Pressing the down arrow several times displays the remainder of the output. We are 95% confident an individual student with a 75 midterm grade and 2 absences will earn a final grade between 75.3 and 80. The first portion of the coefficients list and the X values used for the intervals are displayed as well.

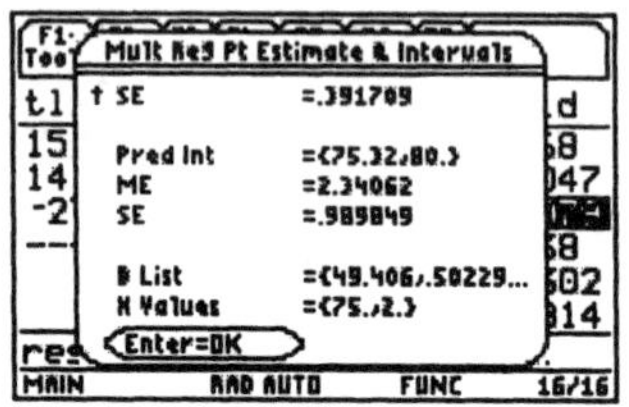

What can go wrong?

Not much that hasn't already been discussed. The most common errors are misspecification of lists and having more than one plot "turned on" at a time.

How do I get rid of those extra lists?

From the Statistics list editor, press [F1] (Tools) and select option `3:Setup Editor`. Performing this action recovers any deleted lists and deletes any calculator-generated lists.

The Complete Internet Marketing Strategy Guide

By Executive Business Coach Ed McDonough